Nature Is Never Silent

how animals and plants communicate with each other

MADLEN ZIEGE

Translated by Alexandra Roesch

SCRIBE
Melbourne • London

Scribe Publications
2 John Street, Clerkenwell, London, WC1N 2ES, United Kingdom
18–20 Edward St, Brunswick, Victoria 3056, Australia
3754 Pleasant Ave, Suite 100, Minneapolis, Minnesota 55409, USA

First published in German as *Kein Schweigen im Walde: Wie Tiere und Pflanzen miteinander kommunizieren* by Piper 2020
Published in English by Scribe 2021

With 33 illustrations by the author.

Typeset in Portrait Text by the publishers

Printed and bound in the UK by CPI Group (UK) Ltd, Croydon CR0 4YY

Scribe is committed to the sustainable use of natural resources and the use of paper products made responsibly from those resources.

978 1 913348 24 3 (UK edition)
978 1 922310 13 2 (Australian edition)
978 1 950354 81 8 (US edition)
978 1 922586 04 9 (ebook)

Catalogue records for this book are available from the National Library of Australia and the British Library.

scribepublications.co.uk
scribepublications.com.au
scribepublications.com

Contents

Introduction

Every living organism communicates

Who have you communicated with so far today? With your partner, your pet, or your plant? And while we are on the subject: how and why did you communicate? The psychotherapist and communications expert Paul Watzlawick hit the nail on the head when he said: 'You can't not communicate.' So it's no wonder that we are constantly exchanging information with other people — within our family, with friends, and with colleagues. But what about all the other living organisms on our planet? Does Paul Watzlawick's 'you' also apply to animals, plants, and bacteria — can these also not 'not communicate'? This book is about something called 'biocommunication'. Every living thing actively sends and receives information and is therefore able to communicate! So 'bio', from the Greek root word *βίος/bíos*, simply means 'life'. Communication, from the Latin word *commūnicātiō*, means 'to share'. 'Bio' fits 'communication' perfectly because life needs to send and receive messages to live. And so even the living organisms in the stillest forest, from the smallest fungus right up to the biggest tree, have quite a bit to say to each other. Those who think the forest is silent just haven't listened properly yet. Nature is never silent.

Why do we need this book?

NATURE IS AMAZING

My enthusiasm for biocommunication began in the woods, meadows, and waters of my hometown in the German state of Brandenburg. Here, everything around me chirps, moos, or cackles, and I trained myself early on to communicate with my fellow living organisms. The many fairytales, myths, and legends in my favourite books proved to be true: here, people could talk to animals and plants; here, the wisdom of nature could help little heroes such as myself out of any hopeless situation. Today, I know that in earlier cultures — the Celts, for example — it was perfectly natural to communicate with nature. To this day, some inhabitants of Iceland and Ireland still ask 'Mother Nature' for permission when new building projects are pending. The indigenous Ainu people on Hokkaido, the northernmost of Japan's main islands, also regularly speak to animals and plants, to strengthen their own connection to nature. Why would people seek to have a conversation with nature if they weren't expecting an answer?

WHAT DO FISH HAVE TO SAY TO EACH OTHER?

I studied biology at the University of Potsdam and quickly worked out what I wanted to do with my life — I wanted to become a behavioural biologist. I wanted to learn everything about why animals behave the way they do and, most importantly, how and why they interact with each other. I was especially interested in cats, and so it was my goal to investigate the communication behaviour of these mysterious animals. As so often happens in life, things turned out a little differently than I expected, and I ended up in Mexico during my final thesis — with no cats at all. My first research project was on fish. I was not particularly happy with this development in my career as a behavioural scientist because, in my opinion, fish were not exactly the most exciting research subjects in terms of communications. However, 'my' fish were different!

The Atlantic carp, *Poecilia mexicana*, and the Grijalva mosquitofish, *Heterophallus milleri*, belong to the fish family of livebearers, whose members lead a dissolute sex life. Most fish don't really have much to do with the opposite sex, because they practise external fertilisation: the female lays the eggs, the male swims over them, the deed is done! Live-bearing fish like the Atlantic carp or the Grijalva mosquitofish, on the other hand, have internal fertilisation. Here, the sperm of the male must somehow enter the body of the female to fuse there with the egg. Clearly, this form of fertilisation involves much more communication between the sexes! As if the 'dialogue' between males and females is not already challenging enough, shoal-living fish are part of a large communication network. This means that male and female fish are rarely alone in order to communicate with each other without being disturbed. The information the two 'lovers' send each other can also be picked up by others in the shoal, and there is always the odd voyeur or two – or rather, eavesdroppers. It was precisely these love triangles within communication that I was interested in for my diploma thesis. I conducted behavioural experiments: for example, to find out if males behave the same way if there is another male present or not. Are they interested in the same females, or do they change their flirting strategy? You'll find the answer to that question in this book!

The Atlantic carp (*Poecilia mexicana*) belongs to the viviparous fish. The male (left) selects a female (right) and fertilises them internally.

TOWN AND COUNTRY RABBITS HAVE DIFFERENT TOPICS OF CONVERSATION

My fascination with the exchange of information in nature continued after my diploma thesis, and it was still my dream to research the communication behaviour of cats. In May 2010, I came to Goethe University in Frankfurt to talk to my future doctoral supervisor about a research project on communication in cats. Again, things didn't turn out as planned. That same night, I was out on the streets of Frankfurt on a bike with no lights when it happened: an inexperienced young rabbit suddenly hopped onto the cycle path. At the last second, I was able to avoid a head-on collision with the rabbit by crashing into a hedge barrier at the side of the path. The rabbit and I came out of the situation with only a few bruises and a big fright. However, I did wonder what this wild creature was doing on the streets of Frankfurt. The next day, my supervisor asked me about my bruises, and I told him about the unusual collision in the middle of the financial metropolis. 'I've always wanted to research wild rabbits,' was his reply. He suggested I complete my doctorate on the communication behaviour of these small long-eared creatures. I stubbornly tried to convince him that cats are much more exciting and that they were the real reason why I wanted to be a behavioural biologist in the first place. He wouldn't drop the subject and so I gave Frankfurt's wild rabbits a chance.

I studied the research and sat down in the park to observe the animals more closely. To my surprise, wild rabbits have a very special type of communication: they use communal excrement and urine points – latrines. These are the means of communication for many mammals that live in groups. Even more interesting to me was the fact that wild rabbits seemed to feel very comfortable in the middle of Frankfurt. To the delight of tourists, the animals would sit in front of the opera house or the skyscrapers of the stock exchange. I found this a really odd sight, and I wondered what on earth could be attracting wild rabbits to the German financial centre: was it the richly laid table through every season, the warmer temperatures of the city, or the many hiding places in the dense vegetation? From studies on birds, I knew that the communication behaviour of animals can change in the city. So I conducted a comparative study between rural and urban rabbits using the latrines to find out how their communication behaviour differs. Do town and country rabbits 'talk' about different things and therefore arrange their latrines differently? I promise you that we'll get to the bottom of this question!

AND WHAT DOES ALL THIS HAVE TO DO WITH US HUMANS?

The more I studied biocommunication, the more I realised that my own communication skills are not the best: I often don't listen properly, I sometimes avoid questions, and I can be unsure what it is I actually want to say. But what some people might consider excellent communication skills, others consider verbal overkill. Coming from the state of Brandenburg, it's a lot for me to even utter a monosyllabic 'morning' at the start of the day. When I was doing my doctoral thesis at the Goethe University in Hessen, my colleagues found this a bit odd. They greeted me with at least four more words each morning. During a visit to Stuttgart, I saw that even this could be topped. There, the morning greeting can be up to ten words long. This was definitely too much for my morning communication capacity. So does this mean that someone from Swabia is more communicative than someone from Hesse or Brandenburg? What is the communication optimum?

In my search for answers to these questions, I took part in numerous courses and events on communication: from science communication to 'elevator pitch' training, all the way to science slams. Alongside my work as a behavioural biologist in the field and in the lab, I was my own research subject. I came into contact with many people and told them about my research and the daily problems of human communication. Those I spoke to were fascinated by the concept of the complicated latrine patterns of the rabbits, which for animals are almost the equivalent of our social media. I kept being asked how communication in nature works and whether plants and bacteria also communicate. What is nature's secret for functioning communications? How can we humans benefit from this in our everyday lives? I began to address these questions more and more and came across the most fascinating connections. In this book, I combine my knowledge as a behavioural scientist with my own experiences of everyday communication, to answer these and other questions.

Life's to-do list

Before we dive deep into the world of biocommunication, we first need some theoretical background knowledge. We know that 'bio' means life – but what is 'life' actually? Which characteristics are common to all living organisms, and how many of them are required for life to call itself 'life'? Generations of scientists have racked their brains over these fundamental questions, and no definitive conclusions have been drawn on this subject. What we know at this point in time is that there are some characteristics, like reproduction or the ability to respond to the environment, that we recognise as belonging to life. Now it is time to take a closer look behind the scenes of life — I hope you enjoy it!

LIFE KEEPS ORDER

The German saying 'Order is half of life' should actually be worded 'Order is the whole of life' because without order and structure, there is no life in this world. Order shows up at all levels and ensures that everything has its place and isn't randomly buzzing around. Atoms are building blocks that can combine into molecules. Molecules in turn can be organised into the components of a cell. The word 'cell' comes from the Latin *cellula* and means something like 'small chamber'. So a cell is sealed off from the outside world by a solid wall or flexible membrane. In the small chamber, there is everything that is needed for life. Many such cells make up multicellular living organisms such as animals and plants, and the principle of organisation and structure is found here too: some cells are responsible for metabolism, others for movement, still others for the transmission of information. All cells with the same task belong to a union of cells known as 'tissue'. Tissue with the same function belongs to an organ. Organs with similar tasks in turn form an organ system, each with its own function, such as the supply of food and oxygen to cells performed by the cardiovascular system. If there were no order in the small things — for example, in the arrangement of the cells — then there would be no order in the big things — such as the symmetrical form of a blossom.

LIFE TRANSFORMS SUBSTANCES

We humans know only too well from our everyday lives how quickly order becomes disorder. Energy is needed to keep everything in place and to maintain order. When you clean and tidy your home, the energy for the vacuum cleaner comes from the wall socket. Unlike this household appliance, you are a living organism, and you can't just get your energy from the wall. So energy can mean different things. To you, me, and every other living creature, chemical energy is crucial to the preservation of order. This energy is contained in the food that every living creature consumes. So the exchange of nutrients is therefore an additional characteristic of life: metabolism maintains the order of the cells and therefore maintains the whole living organism. Without energy from food, life can't receive or transmit information, and communication can't take place.

LIFE PERCEIVES ITS SURROUNDINGS AND REACTS TO THEM

In its entirety, an ecosystem such as a forest is a unique composition of all living and non-living components of the environment. These non-living elements include every grain of sand, every cubic metre of air, and every drop of water! An earthworm can perceive a stone on the ground and, if necessary, find an alternative route. The inanimate stone, however, gives no reaction to the earthworm. A defining characteristic of all living organisms is the ability to perceive their habitat with the help of receiver systems and then to react. Thus, the habitat is full of visual, auditory, chemical, and electrical data. This data only becomes information when a living organism is able to perceive it with its receiver cells. Such receiver cells are also called 'receptors', from the Latin word *receptor,* which means 'sensor'. The type of receptors decides which information a living organism perceives: the animal sensory organs 'eyes' are made for perceiving colours and shapes; 'noses' are perfect for perceiving smells. Receptors therefore enable a living organism to find its way in its own habitat: where is the light or the water, and where can I move to without bumping into a stone? If one organism encounters another organism, then they can both receive and exchange information by means of their receptors. The ability to exchange information is in turn the basis for communication! Only the exchange of information by living organisms among themselves and their consequent interaction with their inanimate environment results in the creation of the big picture, the ecosystem.

LIFE MULTIPLIES

Omnis cellula e cellula. This melodious Latin phrase means: 'Every cell emerges from another cell.' Life reproduces itself and thus passes on its own blueprint, its DNA, to its descendants. In the best case, these offspring are able to reproduce again. Reproduction doesn't necessarily have anything to do with sex. A single cell can double by dividing itself and therefore multiply. This multiplication by cell division is also called 'asexual reproduction' and is mainly found in unicellular living things like bacteria. The cell multiplies its cell components, including its own construction plan, and divides. Under favourable conditions, some bacterial species can double every ten to 20 minutes and therefore bring forth two identical daughter cells. Asexual reproduction is also called 'monogenous reproduction' because it gets by without sexes such as 'male' and 'female'. For an asexual reproducing creature, the elaborate search for the opposite sex is unnecessary.

Sexual reproduction is quite different: here, the gametes of two similar living organisms merge with each other. The special feature of these cells is that they bring a halved DNA construction plan with them. Only the fusion of the gametes to form a common cell completes the construction plan again. Starting with this cell fusion ('fertilisation', which produces a cell called the 'zygote'), a new organism can develop from cell division. The offspring created by sexual reproduction differ both among themselves and from their parents. Sexual reproduction is usually employed by multicellular organisms, such as fungi, plants, and animals. The gametes are not always male and female: organisms such as fungi can have several thousand different sexes!

LIFE GROWS AND MOVES

If the fertilisation of a multicellular organism is successful, the new life can grow and thereby increase in mass. This mass is based on the division and stretching of the cells. The more cells divide and stretch, the more growth takes place on other organisational levels, such as tissues, organs, etc. – this applies to the circumference of trees as well as the circumference of tummies. How much variation there is in terms of growth in nature can be seen in the following extremes: one of the largest known living organisms is the underground-growing mushroom *Armillaria ostoyae*. One specimen covers an area of around 2,385 hectares in an American nature reserve in Oregon – more than 1,700 football pitches. Experts estimate that this fungus is an impressive 2,400 years old. In contrast, one of the smallest living organisms is only a tiny 350–500 nanometres (less than 0.0000005 metres) in diameter and carries the name *Nanoarchaeum equitans*. Translated from the Latin, this means something like 'ancient dwarf that rides on the fireball'. This single-celled organism was not given its name by chance. In fact, the ancient dwarf does actually ride around on the back of another single-celled organism, called *Ignicoccus hospitalis*

– also known as a 'fireball'. While we're on the topic of riding around: the ability to move is another characteristic of life, even in fungi and plants that seem immobile at first glance.

LIFE CONTINUES TO EVOLVE

The face of our planet has changed a great deal over the last few millions of years and with it the living conditions that prevail on it. Sometimes it was hot, sometimes cold, sometimes there were a lot of nutrients and at other times less. But life has never stopped and has always adapted to the new conditions. To do this, it had to evolve, and it was precisely this ability to further develop that is the final characteristic of life. So a cell can manage quite well on its own, but it is only in combination with other cells that it is able to take on new tasks. We can imagine the development of multicellular fungi, plants, and animals like building a house: if we place individual bricks together properly, this results in a house; the house is more than a pile of bricks – it has a completely new function. Multicellular organisms are made up of individual cells and can do more than the individual cell or the simple sum of individual cells. Just like in a real house, we find the principle of organisation and structure in the individual rooms of multicellular organisms too. A house is divided up into different rooms that, in the way they are furnished, perform very specific tasks – for example, the kitchen for food preparation. When life moved from water to land, this new habitat demanded innovations such as complex systems for regulating hydration.

A world of information

Now let's turn to the second part of 'biocommunication' and so to the question: what is communication? In the course of my research and discussions with scientists from other fields, I have come across several definitions and theoretical models. The answer to this question could undoubtedly fill the rest of this book, because communication is an entire world of its own, with countless aspects. If we ask a psychologist, they will give a different answer than a computer scientist or communication scientist. Even among biologists, there is an ongoing debate about when a living being actually begins to interact with another.

HOW DATA BECOMES INFORMATION

The term biocommunication can be summarised as 'the active transmission of information between living beings'[1] – so far, so good. But it leads us to two new questions: what is information, and how can living beings actively send it? Although this seems quite straightforward at first glance, the word 'information' is actually complicated and has led me into some lengthy discussions. If a human interprets data, the data provides useful information to that person. However, 'interpretation' already assumes that the data has been perceived. This is where the receptors come in.

In my opinion, the idea of reading a newspaper clarifies the difference between data and information. Only when you read a newspaper do you perceive the data in the form of letters, words, and entire sentences. If you interpret this data correctly, then the informational content of the newspaper becomes available to you. One prerequisite is that you speak the same language as the people who put the paper together. Bacteria, fungi, plants, and animals

1 Tembrock 2003, p. 10.

are also constantly surrounded by data in their habitat. The data from a forest, lake, or meadow originate from the properties of its components. Besides all the living things, inanimate things like water, stones, and light are also factors. Each of these components has measurable properties that distinguish them from each other. A bird looks different, sounds different, and smells different from a tree or a stone. This means that data in nature, such as colour, shape, sound, or smell, become information only when living beings perceive it through their receptors.

SIGNALS – MESSAGE RECEIVED AND UNDERSTOOD

The way the active transmission of information within communication works can be explained by a simple model. In the USA in the 1940s, mathematicians Claude E. Shannon and Warren Weaver developed this model based on human communication by telephone. With the help of a transmitting device, the telephone, the person who is the transmitter packages the data into a signal. As soon as the transmitter calls the receiver, who has a receiving device, also a telephone, then the signal can be sent. When the receiver perceives the data packaged in the signal, this data becomes information again. If a living being wants to send active information to another living being, then it can send this packaged as a signal so as to improve transmission. 'Packaged' means that, depending on the reason for the communication, certain types of information can be combined with each other. In this way, many different signals are generated – for example, danger warnings to other members of the species.

Let's use another example to illustrate the whole thing: if a male blackbird is in a state of heightened sexual excitement and wants to persuade a female blackbird to mate, it packages this information in an acoustic signal called a 'courtship warble'. This warble consists of a sequence of tones in a certain pitch. In addition to this acoustic signal, the blackbird also sends visual signals to give his motivation to mate further emphasis. Such visual signals might be certain postures or movements. In the case of a blackbird, these include the trembling of slightly drooping wings. The available light, air, and water in the blackbird's habitat are the channels for transmission of such signals. A female blackbird in the vicinity can not only pick up the visual and acoustic signals sent by the male with her eyes and ears, she also recognises the information content of these signals and can thereby gauge the motivation level of the male to mate with her. Now it is up to her to react to the signal and reply to the male blackbird's enquiry of 'Do you want to mate with me?' with a 'Yes', 'No', or 'Maybe'.

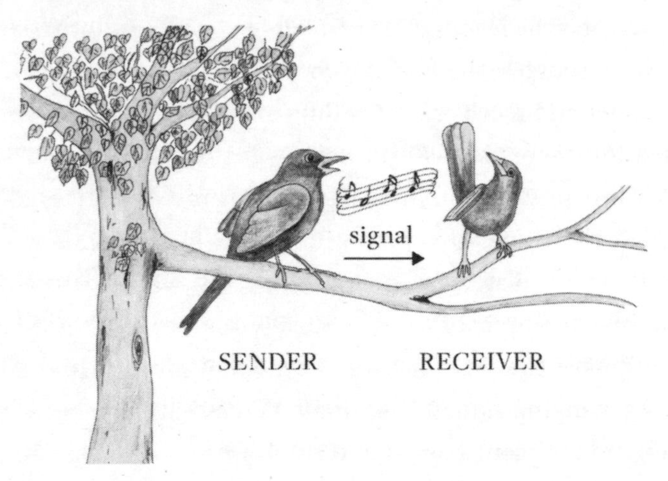

Communication according to the Shannon–Weaver communication model. The transmitter (male blackbird on the left) uses his transmitting device (courtship warble) to send a message through the channel to the receiver (the female blackbird, right). The receiver can use its receptors to unravel the information packaged in the signal.

WHY COMMUNICATE AT ALL?

How does the female blackbird know that the signals 'loud screeches' and 'trembling wings' are meant for her and that this show means that a male of her species wants to mate with her? When it comes to something as fundamental as procreation, the recognition and interpretation of communication signals are mostly inborn. The same sequence of information already served the parents of the two blackbirds as a signal for reproduction, and their parents before them. But the meaning of many signals is learned, when the offspring observe their parents' and siblings' behaviour patterns, thereby discovering which signals are important for their own communication. The emergence of such communication signals over many generations can be explained through a mutual exchange of information between sender and receiver.

The active sending of information involves an effort on the part of the sender, and the reaction of the recipient to this information also requires resources: this is only worth it if there's a benefit – for both sender and receiver. Depending on who the information is intended for, there are diverse motives for communication in nature. A win-win situation arises if sender and receiver benefit equally from the result of the communication. Between living creatures that are related to one another, such as parents and their offspring, the probability is especially high that the sender and receiver have the same reasons for communicating and are therefore exchanging honest information for mutual benefit. If the sender and recipient have different interests in the outcome of the communication, it's not uncommon for nature to send false messages. Signals can then contain information that doesn't correspond with the actual characteristics of the sender – for instance, making him seem larger than he actually is. As you will see later on in the book, such conflicts of interest occur especially between the sexes.

EAVESDROPPING AND TAP-PROOF CHANNELS

Let's get back to our blackbirds. The conversation between male and female blackbirds does not take place in secret, but on a 'public channel' within their habitat. There are many other creatures here that can also perceive their environment by using their receptors. For example, a cat has receptors that help it to perceive the chirping of the blackbirds and thereby listen in to their communication. However, the courtship warble of the male blackbird does not have the same effect on a cat as it does on a female blackbird. For a cat, the perceived information says something along the lines of: 'Here's an easy dinner!' By listening in on the bird communication, the cat gains information that it can now use to its own advantage. Knowing the location of its prey, it can sneak up on the blackbirds. In the worst-case scenario, the communication between male and female blackbirds will end in the killing of the male by the cat.

If the blackbirds see or hear the attacker, then the perceived information with content 'cat coming' is a stimulus to the birds. The male blackbird would then emit a warning cry, which clearly distinguishes itself from the courtship warble in its tone and sequence. The acoustic signal 'enough now, threat of danger' is also recognised by the female blackbird, and she hastens to safety. For the cat, the warning call has a different meaning – its presence has been discovered.

Many animals of prey are well aware that communication made in public can be used against them by predators. Often, secure communication signals are transmitted via 'private channels' to guard against these spy attacks. A lot of insects use optical signals in the range of ultraviolet light to communicate with fellow members of the species; their predators can't detect these, because they lack the corresponding receptors.

LET'S GO!

As you've seen, all living beings send and receive information and therefore have many different ways of communicating with one another. It's particularly interesting to see how living beings interpret the information they receive and how they react to it. This book includes stories about just such information networks between living creatures – I've chosen the ones I'm particularly enthusiastic about to share with you. In the first part of the book, I give you a short overview of how living organisms send and receive information, e.g. can plants hear or mushrooms see? In the second part, we meet nature's transmitters and receivers on land, in water, or in the air. We visit unicellular organisms, fungi, plants, and animals and answer the question: who actually exchanges information with whom and why? I'm talking about genuine friendships between fungi and plants, spying paramecia and lying fish. In part three, I'll tell you all about the wild rabbits in the city of Frankfurt and how information networks in nature interact with the environment of living beings. Once we return from that journey, it's up to you to decide how much of your impressions and the new knowledge about biocommunication you want to take with you into your everyday life. As humans, we are part of life, and therefore there are certainly more parallels for us to discover along the way than we might presume at this moment. Maybe the knowledge about communication in nature will help you when you hit a brick wall during communication exchanges with others – almost like my childhood heroes in fairytales. I hope you enjoy the journey and have a lot of light-bulb moments.

PART I

How is information exchanged?

1

Life is live

So what exactly is this information that a living being sends, and what are the differences between protozoa, fungi, plants, and animals? I will answer these questions in the following chapter, and I bet you'll be amazed at the variety of ways life communicates. To answer the question of how, let's start with the visible – the visual information.

Everything is so nice and colourful here

Our world is full of visual data, and so living beings also use visual information such as colours, shapes, and movements to communicate – from the red-white toadstool, to the shape of an orchid blossom, to the courtship dance of a bird. All this visual information can be used to communicate between living beings of the same species as well as between living beings of different species.

VISUAL INFORMATION — A COMMUNICATION BARGAIN AMONG THE SIGNALS

If the transmitter and receiver remain within sight of each other, then the transmission of visual information is a 'communication bargain'. Information can be exchanged quickly and economically and with minimal loss in the form of colours, shapes, and movements conveyed by means of visual signals. Yet colours and shapes are not particularly flexible methods of communication. We humans can dye our hair, put make-up on our faces, or change our clothes — and in this way send new visual information every day. Most living things, apart from a chameleon or a squid, can't do this. As far as form is concerned, 'inflatable' body parts like a turkey's snood are the exception.

Living beings like animals can still draw on unlimited resources for visual communication, because they are able to move. Movements of any kind are the most flexible type of visual signal because the transmitter can adapt to a changing communication situation within a short space of time. This is especially important in a rapidly changing environment, like when a creature is surrounded by many fellow members of the same species and needs to adapt the form of the information being sent to each one individually. Movements within communication include entire dances that insects, birds, and fish display on the dance floor. The zigzag-shaped mating dance of the male three-spined stickleback (*Gasterosteus aculeatus*) is probably one of the most famous dance performances in the animal kingdom. That amount of physical effort in communication has its price: depending on the intensity, movement requires a great deal of energy. But it doesn't always have to be a stage-ready dance performance to transmit information.

Among many animals, including us humans, facial expressions play an important role in communication. The 'smile is wiped from our face', or we 'put on a brave face'. Mammals that live in a group with members of the same species, such as wolves and monkeys, have a particularly large repertoire of facial expressions.

However, sending visual information works only if both sender and receiver can see each other. Depending on the habitat and the creature in question, the limits of the field of vision are quickly reached, and so large transmission distances are not a strength of visual signals. A tree quickly becomes an impenetrable obstacle and can mercilessly disrupt the transmission of information. If a female bird can't see the male, then the most colourful plumage and wildest mating dance will be of no use, as the information simply can't get through to the receiver.

Examples of different colouring and patterns among fish. Left: Salvin's cichlid (*Cichlasoma salvini*) has a particularly intense colouring during the breeding season. Middle: female green swordtail (*Xiphophorus hellerii*). These popular aquarium fish develop a red background colour during the breeding process. Right: red-spotted cichlid (*Vieja bifasciata*) with typical dark horizontal lines.

COMMUNICATION CHANNEL FOR VISUAL INFORMATION: ELECTROMAGNETIC ENERGY

Visual information is sent via the communication channel 'light'. But what exactly is light? At first sight, this question seems quite simple and easy to answer. Yet it's actually quite challenging and not just because I'm a biologist rather than a physicist. Harald Lesch, professor for theoretical astrophysics at the Ludwig Maximilian University of Munich, presents the science program *Alpha Centauri*, and, in the episode 'What Is Light?', he gets to the heart of it: 'Light is incredibly fast and has differing energy depending on its wavelength.'

When we humans talk about light in everyday life, we usually mean the daylight that's visible to us. The main source for this visible light on our earth is the sun. The visible light contains the wavelengths of the colours we know. Each colour has its own energy content, depending on the wavelength range: from violet to blue to orange and red, the electromagnetic energy weakens. This energy is also called 'electromagnetic radiation' and is all around us. However, electromagnetic radiation covers a wide spectrum of energy. The range visible to us is only a part of this spectrum. For example, ultraviolet radiation, or UV radiation, lies in front of the violet light that we can perceive and is therefore outside our visual perception. At the other end of the range that's visible to us — so beyond the colour of red — are lower-energy infrared, radio, and microwaves.

PIGMENTS CATCH LIGHT

Anthraquinones, anthocyanins, carotenoids, betalains, or melanins — they sound like a list of unusual girl's names but are groups of pigments from the studio of nature. They're the answer to the question of how fungi, plants, and animals obtain their bright colours. The pigments are usually stored in the surface of a living thing — the skin, hair, or feathers. If the pigments are on the same wavelength as the light, they can trap it or, put differently, absorb it. 'Being on the same wavelength' can also be summarised in another word: resonance. The structure of the pigments determines which part of the visible light they capture and therefore resonate with. Now things get exciting: it's not the energy captured by the pigments that determines the colour! In actual fact, it's the visible light that the pigment can't catch. What happens to this uncaptured light? It's sent back by the pigment, or, to put it more accurately, reflected. It's exactly these reflected wavelengths that give a material its colour. The radiant blue and purple shades of pansy petals are a particularly beautiful example of the pigment group of anthocyanins. They reflect the visible light with the energy content for blue, violet, or red. The carotenoids reflect the energy range for yellow, orange, and red. If all the wavelengths of visible light are absorbed, then living beings see black! Black surfaces 'swallow' all electromagnetic radiation in the visible range. The opposite is the case for white surfaces: the incoming visible light is reflected in its entirety. So white flowers appear to be completely white because they don't contain any pigments that capture electromagnetic radiation. Or put another way: the most components of light are reflected from white surfaces.

Pigments are only half the story when it comes to beautiful colours in nature. The nature of the surface matters too. Many flowers have air pockets from which the incoming light is reflected. A particularly beautiful example is the waterlily *Nymphaea alba*. Here in Brandenburg, where I live, it is seen on many of the lakes, and shines from afar as if dabbed on by a painter. What's the secret of the waterlily's radiant white? In addition to the missing pigments, the waterlily's hydrous flower tissue has the air pockets I just mentioned. When light falls on the tissue, it has to pass through all these air-water layers and is refracted each time. This refraction happens so often that the incoming light is completely reflected: the flower appears white. We humans also encounter this phenomenon of complete light reflection in a snow landscape. The freshly fallen snow shines so brightly because the snow crystals refract the incoming light so many times: the result of this refraction is the complete reflection of light. The nature of their surface also gives some animals their impressive 'bling' effect. Tiny structures on the feathers of a peacock or the surface of a dung beetle refract the light in a special way, making them shimmer.

LIGHT ON — LIGHT OFF: BIOLUMINESCENCE

So how can optical information be sent when there is little or no light in the habitat? Many deep-sea and cave-dwelling creatures become their own source of light in the darkness. In the Waitomo glow-worm caves in New Zealand, I witnessed a very special form of animal communication: bioluminescence. This is the ability of a living being to release energy with the help of chemical reactions and to communicate this energy in the form of light. From unicellular organisms through fungi to fish, there are numerous examples that are able to create bioluminescence and to switch light on and off quickly like a light switch. However, some of them need assistance with their magical glow, such as the deep-sea anglerfishes. They themselves are not able to create the necessary chemical reactions, so they host bioluminescent bacteria as subtenants. The glowing beings in the Waitomo caves don't need any subtenants. These larvae of the fungus gnat *Arachnocampa luminosa* illuminate the pitch-black cave roofs like a starry sky all by themselves.

Nature's orchestra

Grumbling, crackling, growling – let's move from visual to acoustic information. The generation of sounds in nature can easily be compared to the sounds produced by instruments. As in an orchestra, living creatures use different materials to make different sounds. From 'violinists' to 'drummers' to 'wind instruments' – they're all there. But listen for yourself!

ACOUSTIC INFORMATION — THE LONG-DISTANCE RUNNER AMONG THE SIGNALS

The advantage of acoustic signals is that the transmitter doesn't need to see the receiver in order to exchange information. And so some living beings can send out calls so loud that they can be heard several kilometres away. A great example is the call of the male howler monkey. They really do live up to their name, as they can generate calls that resonate up to nearly five kilometres through the jungle with the help of a particularly large larynx and a special bone under their tongue. During my field work in Mexico, I was able to hear for myself how impressive these howlers are.

The disadvantage of this type of communication is its high energy consumption. Anyone who has to use their voice a lot every day knows how tiring it can be to send acoustic information. Mostly, it requires the contraction of muscles, such as the vocal cords, to produce sounds. So the transmitter first needs to create the necessary energy. In addition, making a lot of noise isn't exactly risk-free — especially if the transmitter is right at the bottom of the food chain and therefore a popular source of food for many other living creatures. There will always be a robber or two waiting for its prey to betray its position by sending an all-too-intense acoustic signal.

Another disadvantage of this type of communication is the short service life. No sooner is it sent than a warning or mating call quickly gives way to silence. A female howler monkey might only come into the area when the acoustic signal with the information 'male howler monkey ready to mate' has already faded away. So a key question in the transmission of acoustic information is where the transmitter and receiver actually are. With increasing distance between the two, the time delay increases and thus the likelihood that communication will be disrupted. Especially high-pitched tones, such as those that birds use during their morning concert, quickly disappear into the background noise of the surroundings.

However, the short service life of acoustic information also makes it versatile. A bird that intended to attract a mate can warn off a predator just moments later. Among birds and many mammals, such as the whale, the great variety of acoustic signals is evident in the form of entire songs, with individual verses and melodies.

WHY THERE'S NO 'BOOM' IN SPACE

What is sound? How does it get from sender to receiver? In order to answer this, I would like to take you on a short excursion into the history of film: in the first *Star Wars* film, a space station explodes with a lot of noise in the middle of space. Initially, you might not think too much about this scene, but this may change if you think about the following physical considerations: sound is a mechanical vibration that propagates in a wave-like manner in an elastic medium. In contrast to light, sound is therefore not a form of electromagnetic energy, but the result of oscillating mass particles. These do not necessarily have to be 'solid'. Gases or liquids such as water can serve as a source of sound. Although an attack on a space station with sufficient firepower through laser weapons could certainly cause it to oscillate, there is an important aspect missing from the film scene: the mechanical oscillations create a change in pressure and in the density of the surroundings in which the oscillating matter is located. Only through the presence of a medium can the oscillations continue to propagate as soundwaves. However, space is a vacuum, and therefore the medium needed for the oscillations to propagate as soundwaves is missing. Thinking back to the transmitter–receiver model, the missing medium in space is therefore also the reason why there is no channel there to transmit acoustic information.

THE ART OF STRIKING THE RIGHT NOTE

When we humans speak of sound, we mean all tones, sounds, or noises that we can hear. In this way, we can identify sound sources whose waves are in the range of 16 Hertz to 20,000 Hertz. What exactly does that mean? The unit Hertz (Hz) stands for the number of oscillations in one second, the so-called 'frequency'. If we pluck the string of a guitar, the string begins to swing back and forth. The faster the string swings, the more oscillations there are per second and the higher pitched the resulting sound.

Of course, there's more than just the upper and lower limit of the sound we hear. There are sound sources that send oscillations in infrasound and are therefore below 16 Hertz. In contrast, the frequency at our upper hearing limit is followed by ultrasound at more than 20,000 Hertz. This extremely high-pitched tone can be produced and perceived by bats, for one. If the pitch is determined by the number of oscillations, the volume refers to the size of the oscillations – the amplitude. The wider the oscillation, the louder the tone.

The speed of the soundwaves depends on the properties of the medium, such as its temperature or density. If soundwaves race through a birch tree at 3,800 metres per second, water slows their speed down to 1,450 metres per second, and air at zero degrees Celsius slows them all the way down to 332 metres per second. Figures in decibels (dB), on the other hand, illustrate the different strengths with which a sound event meets its surroundings. But enough of the theory. Let's delve into a sound symphony that only nature could compose. What a pity that we can't hear most of it with our own ears!

WHY THERE ARE 'CLICKS' AT THE ROOT OF A PLANT

Mechanical vibrations generate the movement of individual components of a cell. If there are many cells vibrating on the same wavelength and thus in resonance with each other, they can, like a choir, produce a greater volume than they could alone. Even unicellular bacteria use acoustic waves to stimulate the growth of neighbouring cells. The question of whether the sounds emanating from living organisms actually serve a communication purpose or are just a by-product of everyday biological processes presents scientists with experimental challenges.

Plants also have numerous noise sources — for example, in the vessels for water. Air bubbles are often found in plants trying to survive on little water. When these bubbles come loose, there is a slight popping noise. Australian and Italian scientists listened in (and continue to do so) on the mysterious world of plants and searched for proof that they actively transmit acoustic information for communication with other living beings. They did in fact find something, discovering clicking sounds at the roots of young corn plants (*Zea mays*) with a frequency of around 220 Hertz. These 220 Hertz clicks correspond exactly to the pitch of the sound sources in whose direction the plant roots orientate themselves during their growth. It has been known for decades that plants react to acoustic waves of different pitches. The seeds of cucumbers (*Cucumis sativus*) and rice (*Oryza sativa*) germinate better if they are exposed to pitches around 50 Hertz. And once they've grown to the size of a small plant, the roots grow faster yet when exposed to that 50 Hertz pitch. Even pea plants (*Pisum sativum*) react to the sound of flowing water. So is the corn's clicking noise just a coincidence or a real signal of communication? We can look forward to further results from the plant whisperers.

INSECTS 'FIDDLE' WITH THEIR LEGS AND ARMS

From plants, we now move to animals. You can find all kinds of instruments to produce notes in the animal kingdom, but the principle is always the same: by hitting them, blowing into them, or plucking them, membranes, air columns, or strings begin to vibrate, and this results in a sound. To insects, their legs and wings are what the violin and bow are to a violinist. Grasshoppers, for instance, produce their typical noises with the help of their stridulatory organs. These organs consist of a file and a scraper. Serrations, like on a saw, are located on the inner surface of the femur. The characteristic rasping noise is created during warm summer nights when the hind leg moves across the folded front wing. While both male and female grasshoppers create their sounds in this way, 'fiddling' is a purely male thing in crickets. They also have a file and a scraper, but these are located on the cricket's non-airworthy front wings. With their stridulatory organs, these insects can attain pitches in the ultrasound range, exceeding a frequency of 20,000 Hertz. No violinist can fiddle that fast! Stridulatory organs aren't always needed to produce sounds, though: the beat of a bee's or a beetle's or a bird's wings creates soundwaves. The wings of the smallest mosquito species, for example, beat 1,000 times per second and therefore hit the auditory nerves of us humans.

FROGS ARE WIND PLAYERS

It's not only the humming and chirping of insects that fills balmy summer nights – the croaking of frogs can also be heard through the air from afar. The principle is as simple as it is ingenious and is probably best suited to the wind section of the orchestra. Viewed physically, on a clarinet or a bassoon, it's the column of air that begins to swing as it passes through the closely aligned reeds in the mouthpiece. The vocal cords of many vertebrates, like birds or mammals, perform a similar task: a surge of exhaled air causes these elastic cords to vibrate, thereby oscillating the air. The more strained the cords are, the faster they can oscillate, and the higher the tone produced. Frogs are much smaller than birds or mammals and they struggle to send their acoustic information very far, as the oscillation of their vocal cords isn't strong enough. A sac-like sound bubble on their head serves as an amplifier and gives the frog call the necessary volume.

Equipped with this sound system, a male frog can achieve volumes of between 65 and 90 decibels with his mating call. This comes close to the sound pressure of a pneumatic drill. The next section of our nature orchestra is also very loud, as we are now turning to the percussion instruments.

DRUMMING SPIDERS AND WILD RABBITS

Let's begin with the drummers, who make a membrane vibrate, thereby creating a sound. Insects such as cicadas and some butterflies have drum organs, so-called 'timbals', on their abdomen. These are membranes made of the natural material chitin and are equipped with hard, but moveable, ribs. As soon as the adjacent muscles tense and relax, the ribs start to move, and click noises are produced. Under the cicadas' drum organ sits a sac filled with air, which increases the sounds even further. The drum organs work a bit like when we squeeze an aluminium can in our hand. The resulting buckle then pops back into its original position as soon as we let go. The cicada species *Platypleura capitata* can create up to 390 clicks per second.

Spiders know how to get by without a drum organ and simply use one of their eight legs to stamp the 'beat' on the ground beneath them. The giant crab spider uses its entire body to make leaves vibrate beneath it and thus produces sounds. Mammals such as wild rabbits also use physical effort to communicate acoustically. They are experts when it comes to 'drum rolls'. When there's imminent danger, the animals begin to beat the ground with their powerful hind legs. The resulting soundwaves spread deep into the earth and are a signal for other rabbits not to leave the safe burrow. Rattlesnakes probably have the most famous percussion instrument: hard interconnected plates in the tips of their tails that produce their eponymous sound as the plates rub against each other.

The next musician in our nature orchestra is the gurnard. Members of this fish family live on the seabed and produce their snarling and grunting noises through several 'instruments'. These fish not only rub their hard gill-covers against each other, but also tense the muscles around their swim bladder and thus press air out of it. The air in turn generates soundwaves — a principle that other fish also use to produce sounds. These sounds are mostly rhythmic, and this is also the secret to their information content. Most fish find it hard to distinguish volume underwater, as soundwaves change their speed and spread in comparison to those on land.

Let's stay underwater for a little longer, because this is where we find one of the loudest 'musicians' — or should I say 'gunslingers'? — that nature has to offer.

FINISHING WITH A CRACKER

The big-claw snapping shrimp (*Alpheus heterochaelis*) — 'pistol shrimp' for short — lives in shallow tropical and subtropical waters. The shrimp, which is a crustacean, is only five centimetres long, but it makes a noise underwater that, at up to 210 decibels to the metre, can match the signal of a sperm whale! How can a small crab produce as much noise as a large whale? The secret is in the shrimp's enlarged claws. In both male and female, one of the two claws is much larger than the other and can be up to 2.5 centimetres long.

One half of this giant claw has a cylinder-like depression, while the other half is reminiscent of a piston. By tensing strong muscles, the piston-like half of the claw moves sideways and places itself under extreme tension. This is so great that when the 'piston' snaps back into the 'cylinder', the action produces an extremely fast jet of water – and the loud bang that is typical of the crab. Taking their relative size into account, the sound is comparable in volume to the ignition of a Saturn V rocket, and can't be attributed to the clash of the hard claws. It needs more steam to generate such a powerful wave underwater. The pressure in the claw changes as a result of the fast-flowing water, resulting in a steam bubble, also known as a 'cavitation bubble': the pistol shrimp more or less makes the salt water in its claws evaporate. It's only the collapse of the cavitation bubble at decreasing pressure that generates the impressive bang of 210 decibels. The bangs serve to kill prey like worms or small fish with the resulting pressure wave. But as well as sending, the pistol shrimp also receive: the claws contain hair cells that can perceive the pressure of the water jet of another pistol shrimp. The bangs can be used as something like a warning shot to rivals, as the small crabs don't hold back when it comes to defending their territory. They do well to heed these warnings – a claw flicking directly onto the shell of a rival can have fatal consequences!

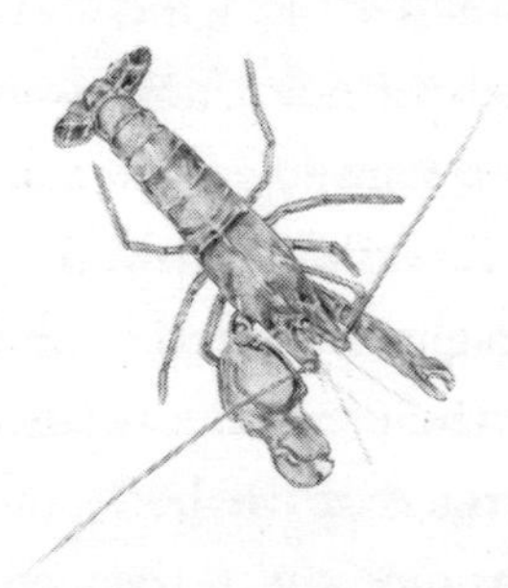

The pistol shrimp, which belongs to the crab family (*Alpheus heterochaelis*) has one enlarged claw, with the help of which it can produce loud bangs underwater.

The world of smells

Now let's enter the realm of chemical information and learn about the oldest form of communication in nature. In this fascinating world of communication, we encounter osmophores, secretions, and latrines and meet creatures that influence the behaviour of others of their kind with their chemical messages. As exciting as this type of information is in nature, we still know far too little about how living beings specifically use chemical signals to communicate – including us humans. Or can you explain why in certain situations you 'turned up your nose', 'kept your nose clean', or 'stuck your nose in the air'?

CHEMICAL INFORMATION – THE PERSISTENT SIGNAL

The advantage of communicating with chemical signals is their wide range. Smells are extremely well suited to passing on information over long distances. Chemical information is easier to produce in comparison to acoustic and also has a longer service life. Like perfume, fragrances or smells 'hang around' for hours after the transmitter has gone. They are, however, not the fastest signals and need time to get from the transmitter to the receiver. The more volatile the chemical messenger substances, the faster they can be blown away by the wind and the further they can spread through channels such as air and water.

Animals and plants send chemical substances as secretions, which can form in special individual cells or entire cell clusters – the glands. These glands may be located inside the body and thereby release their secretions internally. Glands that are located on the surface of the body, such as glandular hairs, release their secretions directly to the outside world. Secretions released externally in the form of gaseous smells, liquid flower nectar, or solid resins serve as an important chemical means of communication.

'Open osmophore!' could be the magic term to elicit the scent from flowering plants such as the orchid or Aaron's rod plants. Osmophores are special glands in the upper cell layer of the petal that hold the precious scent of the flowers like small perfume bottles and release it into the environment. Once they are on their way, the chemical signals need a suitable receiver whose behaviour they can influence. If the communication partner comes from the same animal or plant family, then the chemical messages are called pheromones. If the transmitter and receiver do not belong to the same species, then the transmitted chemical information is called allelochemicals. Plants, for example, send allelochemicals to attract insects for pollination.

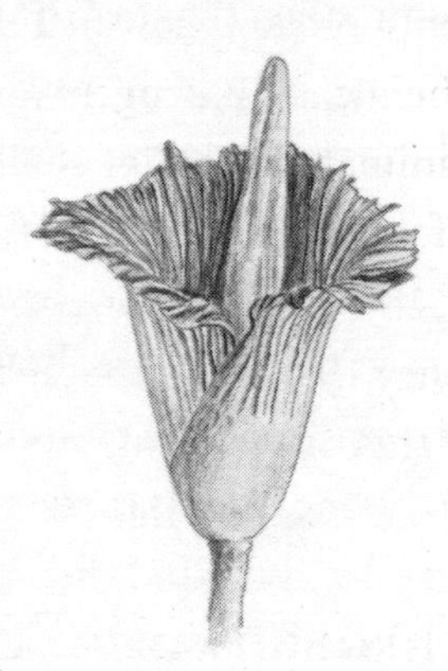

The titan arum (*Amorphophallus titanum*), which belongs to the Aaron's rod plant family, has special scent glands known as 'osmophores' in the upper cell layer of its flowers.

PHEROMONES AND THE COMMUNICATION BETWEEN MEMBERS OF THE SAME SPECIES

Let's stay with the communication between living beings of the same species, which therefore involves pheromones. Even closely related single-celled organisms use these scents to communicate among themselves. The ciliate with the Latin name *Euplotes raikovi* is a particularly communicative example because it emits more than five different pheromone variants. Pheromones are also found in multicellular algae, fungi, and plants – along with optical signals, they are the most important communication signal. Probably the best-known pheromone in insects is the bombykol. It's used by the female silk moth (*Bombyx mori*) to attract males over a distance of several kilometres. The bombykol is so productive that even a single molecule is sufficient to influence a male silk moth's behaviour.

Pheromones should not be confused with hormones, because there is one important difference: in contrast to pheromones, hormones are important messenger substances *within* a living being. The purpose of sex hormones such as testosterone and oestrogen isn't to attract the attention of a fellow member of the species out in the world. These chemical messengers are instead responsible for ensuring that an animal capable of sexual reproduction gets in the mood itself before sending out pheromones to attract a suitable sexual partner. Once the sex hormones have done their job in the animal's body, they leave via the transport vehicles 'faeces' and 'urine'. In this way they unintentionally transmit information about their owner to the outside world.

COMMUNICATION THROUGH FAECES AND URINE

The stuff we humans want to banish from our sight as quickly as possible, thoughtlessly sending it off into the wide world of the sewage system, is the number one method of communication for many creatures in nature. As waste products of metabolism, liquid and solid excrements are conceivably the cheapest and most 'personal' way of communicating, which mammals in particular use to send information. Studies on wild rabbits and badgers, for example, showed that their droppings and urine contained scent that provided individual information about each animal's age, sex, and willingness to mate. To blame for this public data scandal are, among other things, the individual scents that are formed in various glands and added to the faeces or urine. Colour, odour, and quantity of the residues also provide information on the health of their owner.

In vertebrates, urine is produced as a final product during the kidneys' cleaning process of the blood. Like a filter, the kidneys free the blood of everything that should not remain in the body. This includes old blood cells and toxins. Urine therefore consists of waste products dissolved in water, which is transported from the kidney via small urinary canals and collected in the bladder. Once a certain amount of urine has accumulated, pressure sensors are activated, which awakens an urgent need in the animal. The kidneys are also responsible for the body's fluid balance, and, depending on the fluid levels, the level of water in the urine varies.

Faeces, on the other hand, is an end product of the gastrointestinal tract and, among other things, consists of cells from the intestinal mucosa, unused food components, and intestinal bacteria, and products of their fermentation and decay. A functioning digestive system is a sign of physical health for us humans too.

LATRINES — A 'GOOD COMMUNICATION' BUSINESS?

For many mammals who live in groups, excrement and urine play such an important role as a means of communication that these animals don't do their business indiscriminately, wherever they happen to be walking or standing. The repetitive, regular use of the same loo by a species will sooner or later result in poo heaps. These latrines have two decisive advantages from a communication point of view: they are easy to spot and contain concentrated scents of other members of the species. In other words, these places are anything but a quiet spot to sit, and take on roughly the same function as social media for us humans. Wild rabbits exchange information about who is currently searching for a mating partner or who is the highest-ranking male or female in the group. If the remnants are fresh, the predecessor probably only just left and is still nearby. The combination of visuals and scent further enhances the meaningfulness of latrines as a medium of communication – the more often a latrine is used, the more emphasis is placed on both factors. Perhaps this is a small consolation for all those who unerringly walk into every pile of dog shit: it could be a lot worse! Gazelles, rhinoceroses, and the North African hartebeest use latrines that reach several metres in diameter.

WHAT DO ADVERTISING BILLBOARDS AND ANIMAL TOILETS HAVE IN COMMON?

While we humans try to 'do our business' in a hidden place, several animal species – such as rabbits, badgers, and guenons (a type of monkey) – choose particularly conspicuous places for their latrines. Their 'business centres' are often found on raised objects in the landscape or located at open crossroads. The good view from an elevated place is probably also the reason why the swamp rabbit (*Sylvilagus aquaticus*) makes its latrine on tree trunks. The advantage of these eye-catching spots is their good visibility. For communication centres to work, they have to be in places where they can be seen. We can imagine latrines like advertising billboards in the landscape, which need to be strategically placed in a favourable position in order to send the information to the animal. The question 'Where is the toilet?' should not even need to be posed – the location should be obvious.

The North American ringtail cat (*Bassariscus astutus*) uses particularly striking places for its business. This animal is a frequently seen feature in the public parks of Mexico City, where it positions its latrines in clear sight on the blue water pipes. Apparently, it is not only the striking blue colour that tempts the ringtail to place its latrines there. The raised pipes provide the animals with the necessary quiet to do their business away from the hustle and bustle of the Mexican capital. The disadvantage of an easy-to-find latrine is that it can quickly become a death trap. The use of such 'public' toilets carries the danger that popular prey animals such as wild rabbits are served to their enemies on a silver platter. Wild rabbits therefore carefully weigh up the risk of using a latrine against the danger of being eaten by predators. If the risk of being caught by birds of prey or foxes is too high, the animals prefer to place their latrines close to protective vegetation or rather close to their burrow.

2

Life is on stand-by

No information without receptors

Receptors record information from different directions. There are the 'inwardly' directed receptors that gather information about internal processes within living beings. Such internal receptors may react sensitively to the pressure of fluid in cells or blood in the vessels. This is also what happens to the receptors around our stomach, so we know when it is time to stop eating. When it comes to recording information from our surroundings and therefore communicating with other living creatures, the external receptors come into play. The more receptors a living being has, the more it is able to perceive detail about its surroundings. This is how living organisms such as bacteria, which only consist of a single cell, have direct contact with their habitat. Their receptors are situated on the surface of the cell and are directly integrated into the outer membrane. These external receptors react sensitively to light, pressure, or chemical substances. Unicellular organisms like the paramecium are a good example of how life forms that have a very simple structure can perceive their environment and exchange information with it.

You will get to know the paramecium a bit better later in the book. For the time being, it is enough for us to know that it's a unicellular organism that is free-living (i.e. not parasitic), aquatic, and can move around quickly in all directions. During the 18th and 19th centuries, it was colloquially referred to as the 'slipper animalcule', and it is not for nothing that this unicellular organism bears that name: with its long oval shape, the *Paramecium caudatum*, for example, does actually look slipper-shaped (and in contrast to most other unicellular organisms, it can even be seen as a small dot with the naked eye). If a nutrient comes along and docks onto the appropriate receptor on the paramecium's external cell membrane, then the unicellular organism can make its way straight to the food source thanks to this information. The paramecium also gets moving as soon as it perceives danger in its surroundings. There are a lot of dissolved substances in water that do not agree with the paramecium, including carbon dioxide in certain concentrations. In response to such toxins or chemical information from the paramecium's predators, a biochemical chain reaction is set in motion with the aim of moving the unicellular organism away from the source of danger in a targeted and rapid way.

Fungi and plants also 'navigate' their environment by using receptors for visual, chemical, or physical information and can even contact other organisms, in their own fashion – and they don't even need to leave their position! For instance, if the roots of two plants touch, this contact causes a physical pressure to be exerted on the root surfaces. The pressure-sensitive receptors in the root cells perceive this pressure and, in reaction, the plants begin to grow in a different direction, so as to avoid getting into a root turf war.

With more highly developed creatures like animals and humans, particularly large numbers of receptors come together with 'like-minded' cells and form sensory organs such as eyes or ears.

WHAT DO RECEPTORS HAVE IN COMMON WITH DAMS?

Let us think back to the transmitter–receiver model of telecommunications developed by Mr Shannon and Mr Weaver, which I mentioned in the introduction. The receiver's telephone rings as soon as the transmitter has dialled the correct number – provided that all technical components are working perfectly and the line is not busy. Cell receptors also indicate when suitable information arrives. However, in contrast to a telephone, receptors don't ring. In reaction to incoming information, together with the receptor, the cell changes its potential. The word 'potential' is derived from the Latin word *potentia* and means 'strength' or 'power'. The potential offers the power to change the receptor cell, so to speak, as soon as suitable information arrives.

We can imagine the whole thing like a dam, which holds differing amounts of water on either side. If the dam wall were to break, the dammed water would flood forth with all its energy. We humans use dams to generate energy, by allowing some water to flow unhindered and turn a hydroelectric generator – thereby releasing the dam's accumulated potential energy. Conversely, it requires a lot of energy to fill a reservoir with water using a pumping station.

Let us now transfer the image of the dam to a receptor in a living being. The outer membrane of the cell is like the dam wall: it creates the internal and external part of the cell and is impermeable to various substances. There are not only different quantities of chemical substances inside and outside of the cell but also different charged particles. There are particles with positive charges and particles with negative charges — the optimists and pessimists, so to speak. These charges are like the water in the dam and are also 'dammed up'. If a receptor cell is currently inactive and not busy absorbing information, then most of the 'optimists' are located on the outside of the cell; most of the 'pessimists', on the other hand, are inside the cell. Accordingly, on the outside of the cell, the mood is super positive, while the inside is in a pessimistically negative mood. There are gateways in the cell membrane, and, if these gateways in the membrane are open, the charged particles can change sides. But when do the gates open? You probably already have an idea: when the phone rings! The gates in the cell membrane of a receptor cell open as soon as suitable information arrives. Now the electrically charged particles flow to the other side and take their positive or negative charge with them. The more such suitable information reaches the receptor, the more the potential between the interior and exterior of the cell changes. Every living cell has this kind of potential, but only excitable cells like the nerve cells of animals can transmit the changes in potential in the form of action potentials over long distances.

THE LADIES AND GENTS IN THE EXCHANGE

Unlike fungi and plants, most animals cover a lot of ground by changing their location. They need to be able to find their way and constantly take in information from their surroundings. You may know this from your daily life: if you are on the road a lot, your sensory organs receive a lot more information than when you are sitting comfortably at home on the sofa, barely moving. Animals therefore have very special receptors that help them to cope with the daily flood of data and to filter out the most important information. These are the nerve cells.

In their form and function, these cells are completely focused on the reception, processing, and transmission of information. So the 'inbox' is on one side of the nerve cell, because this is where the information arrives at the small finger-shaped protuberances along the cell membrane, the dendrites. The incoming information also changes the potential of the cell membrane. The more information that arrives, the more charged particles can change their place, but only if the information is flagged as 'important'. If the incoming information is flagged as 'unimportant', then it will never get to the nerve cell's 'outbox'. This is located on the other side of the nerve cell and is also a protrusion of the cell body. This bulge is much longer than the dendrites and is called the 'axon'. We can imagine the axon as an ethernet cable that carries information to the inbox of another cell. This axon can only receive one or zero bits of information – the nerve cell doesn't have more possibilities of expression. The transmission of information by the axon works roughly like our Morse code. The actual information is hidden in the frequency and the distance between the electrical signals, not in the strength of the signal. The transmission of information in the axon only takes place when the gates in its cell membrane are open long enough and almost all charged particles have changed sides. The complete change of the potential on the axon of a nerve cell is also called 'action potential'. The mood here is positive on the inside of the cell and negative on the outside, or, to put it differently, the water is no longer dammed up and its full potential is unfolding. The opening of the gates in turn depends on how much important information has landed in the inbox with the dendrites. In this way, the potential at the axon has to be recreated again and again so that it is available to pass on information. Like our dam, it also needs energy-operated pumps. This way, the electrical excitation propagates along the axon. The ends of the axons are very close to the dendrites of another nerve cell: only a wafer-thin gap separates the two cells from one another. Chemical messengers now transfer the information from the end of the axon across the gap to the dendrites of the next nerve cell. The more often the axon sends a Morse-code message, the more messenger substances set out on their way, and the more charged particles can change places on the cell membrane of the dendrites.

NERVE CELLS CONNECT TO NERVOUS SYSTEMS

Nerve cells are linked to other nerve cells to form nervous systems, and these nervous systems ultimately inform other cells in the body by being directly or indirectly interconnected with them. Nerve cells include those that can trigger an effect in living organisms – for example, in the muscle cells.

Depending on the stage of development of an animal, the number of nerve cells and thus the size of the nervous system differs. Coelenterates like the sponges or cnidarians in the sea are very simple creatures with only a few individual nerve cells, connected to the simplest form of nervous system: the nerve network. For many invertebrates, such as snails, insects, and spiders, there is an accumulation of nerve cells in the head and abdomen. An accumulation of nerve cells in the head area is necessary to smell, see, or hear; in order to get the image, sound, or smell, a nervous system with a central switch point like the brain is a prerequisite. Most nerve cells are found in the central nervous system. In vertebrates, the central brain and spinal cord are safely protected by the skull and the spine.

The receiving systems for information on the surface of a living being, the receptors, can't hear, see, or smell for themselves. They simply 'translate' the information into a common language and then pass it on. Only the brain can correlate all incoming information from the various receptors with each other, relate it to memories, and, if necessary, initiate appropriate behavioural responses. At the centre of the relatively new field of research of plant neurobiology is the question of whether plants actually have nerve cells and therefore a prerequisite for a brain. The presence of electrical signals or chemical messenger substances such as dopamine or serotonin provides an indication that there's far more going on in our green friends than we previously assumed – even without the presence of an actual brain.

WHAT'S THE POINT OF EYES WHEN YOU LIVE IN THE DARK?

The aim of life is to stay alive, and communication is also about life and death. If a living being wants to communicate with another living being, both have to go into resonance with each other – in other words, speak the 'same' language. The transmitter must be sure that the receiver has the appropriate hardware and software for the information being sent. Put another way: if you want to call someone, you need the number of the receiver and the receiver needs a telephone. Using the example of the cave fish, I would like to show you to what extent a habitat determines whether a number is allocated or not.

I already introduced you to one of my study subjects in the Mexican jungle, the Atlantic carp (*Poecilia mexicana*). This fish species is found both outside the cave under daylight conditions and in absolute darkness in the deepest chambers of the cave. Male Atlantic carp in open water have a distinct orange colouration on their fins and are therefore easy to distinguish from the less-colourful females. Their fellow members of the species in the cave lack this colouring, and the proverb 'All cats are grey by night' rings true here. In addition to the lack of colour, the eyes of the cave carp have regressed strongly, thereby limiting their function.

Its pale colouring and reduced eyesight make the cave fish seem like a ghost guarding the underground halls. It is an impressive example of how economical nature is: what is not needed is simply not produced in the first place – or if the circumstances change, is rationalised. Why invest time and energy in training eyes if the channel 'visible light' can't be used for communication anyway? It makes no sense to buy an expensive phone if there's no reception in your town.

Examples of the Atlantic carp (*Poecilia mexicana*) live in the darkness of Mexican limestone caves with reduced eyesight.

Here's looking at you, kid

In the film *Casablanca*, Humphrey Bogart whispers the famous line 'Here's looking at you, kid' into the ear of his love interest, Ingrid Bergman. If Mr Bogart had been a pedantic biologist without a sense of romance, he would probably have chosen the words 'Using my light sensory organ on you, kid'. I doubt it would have had the same romantic effect. Animals' eyes are sensory organs for light, and at their core are the light-sensing cells. These sensory cells are special nerve cells with the ability to capture light via a chemical pigment. If the exposure of the pigment changes, the electrical potential of the sensory cell also changes. If the pigment reacts to light in the visible range, then the light-sensing cells are called photoreceptors, i.e. 'light receivers' – but you'd best see for yourselves!

LIGHT-SENSING ORGANS CAPTURE ELECTROMAGNETIC ENERGY

'The forest has eyes.' This newspaper headline caught my attention on a train recently, and I immediately pictured scenes from a fantasy novel in which trees had eyes and could see like us humans. In most cases, such magic trees also display an atypical desire for movement and leave the forest to explore their surroundings. I carried on reading and stopped after three sentences: the article wasn't about trees wandering around the place with open eyes. It was, rather, a report about hunters who are increasingly putting up cameras in the forests, hoping for a snapshot of foxes, wild boar, or even wild cats. So it's not the trees that we need to feel watched by during our next forest walk – or maybe it is?

In a way, plants are at the forefront of 'seeing' – after all, they possess many receptors for incoming light. With the chemical pigments present in leaves or flowers, they capture a wide range of electromagnetic energy. The leaves are green because the pigments inside them absorb the red and blue wavelengths of visible light but not the wavelength range for green. Plants can react then and there to the received information by using the light-sensing cells in their leaves. These receptors use the incoming electromagnetic radiation to measure the number of hours of sunshine or the time – for example, the proportion of red and blue light over the course of the day. If the sun is low in the sky in the morning and evening, a large amount of red light hits the earth. The proportion of blue light, on the other hand, is greatest around midday, when the sun has reached its highest point and its rays are perpendicular to the earth. When large amounts of blue light hit the plants' receivers, a shielding process takes place whereby the leaves turn away. The question of flowering or not-flowering is also decided by the incidence of the sun's radiation. So-called long-day plants like the forest tobacco (*Nicotiana sylvestris*) only flower when the length of a day is more than 11 hours. Many aquatic unicellular organisms, such as representatives of the group *Euglena*, also practise photosynthesis. They have a red pigment on their surface and with this 'eye spot' are able to turn towards the light.

Creatures that have a basic structure, such as worms, have few light-sensing cells in one place and can only recognise which direction the light is coming from and how intense it is. In more highly developed animals, the brain's evaluation of the information received by many light-sensing cells is what enables them to perceive their surroundings. In particular, vertebrates that live on land, including us humans, need a good idea of what their environment actually looks like. What shape and colour are other creatures, where are they, and how quickly are they moving? Efficient eyes therefore consist of many light-sensing cells and are real organs, made up of different tissues. These include a lens, which pools incoming light on the light-sensing cells.

PIGMENT, CUP, AND CELL — HEY PRESTO, A FLATWORM'S EYE

How a living creature sees its surroundings depends on many factors. The location and number of light-sensing cells are important components and are adapted to the animal's particular habitat. Turbellarians, which belong to the flatworm family, have to be able to find their way around their habitat – which is water – because they are a popular snack for several predators that might swim by. And so flatworms such as *Dugesia tigrina* have very simple eyes, known as 'pigment cup eyes', which are cup-shaped with a single type of photoreceptor.

The photoreceptor is made up of a pigment, a cup, and cells. The cup-shaped cell layer, which contains light-sensing cells, is covered by pigment like a blanket. With the exception of a small aperture, the pigment blocks the light receptors. What is the point of blocking, if the light-sensing cells are there to capture the optical information? This is where the small opening that is not covered by the pigment comes into play. On one eye, this aperture is on the left side at the front of the turbellarian's head, while on the other eye, the opening is on the right side at the front. The light-sensing cells conduct information about the angle and intensity of light incidence to the accumulation of nerve cells in the worm's head. These nerve cells in turn give the impulse 'turn head' until the incidence is equally low in both eyes. Not being in the spotlight and shying away from the light is vital for the little flatworm. This is how it manages to protect itself from predators, by hiding in the dark corners of its habitat and under stones.

Speaking of dark corners: have you ever had the experience of waking up in the middle of the night, drowsily stumbling into the bathroom, and switching on the light, only to see a small silvery something slip down the plughole?

Flatworms such as *Dugesia tigrina* have very simple eyes that are cup-shaped with a single type of photoreceptor.

INSECTS AND CRABS SEE IN MANY DIFFERENT WAYS

It's not only silverfish that like to tour damp bathroom areas, but all kinds of insect. Silverfish possess a head, with sense organs and mouth parts; a chest, with three pairs of legs; and one abdomen, containing the digestive tract and reproductive organs. Our initial focus is on the insect's head because this is where the eyes with their light-sensing cells can be found.

For most arthropods, like insects and crabs, the issue of 'seeing' is a complex matter. Complex because they have compound eyes, made up of several thousand individual eyes. Each individual eye is called an 'ommatidium'. It's immobile and only ever picks up light from the angle it's facing. The part of the ommatidium facing outwards has a transparent lens that allows incoming light to pass to the light-sensing cells below. The light-sensing cells in turn contain the light-absorbing pigment rhodopsin. Also known by the pretty name 'visual purple', the pigment is a universal 'light trap' and is also found in the eyes of vertebrates. In the compound eye of arthropods, the axon on the lower end of the ommatidium passes the information about the light exposure of the rhodopsin to the brain. In this way, each mini-eye captures one focal point and therefore one single facet of the surroundings. The differences in light incidence on the individual ommatidia are put together to form a mosaic picture in the arthropod's brain. This mosaic picture is not particularly detailed compared with human vision, but it renews itself up to 300 times a second, six times more often than in us humans.

The compound eyes of dragonflies are especially impressive, and some species feature several tens of thousands of ommatidia per compound eye, taking up almost the entire surface of the head. Like most insects, dragonflies are on the menu of several predators. Their survival depends on them being able to spot their predators coming very quickly! On the other hand, dragonflies are excellent hunters themselves and are able to catch their prey in flight. Neurobiologists at the University of Arizona took a closer look at an example of the long-tailed skimmer (*Plathemis lydia*), which hunts flies. With special markers on the body of the dragonfly and a camera that takes 200 pictures per second, the scientists followed the head and body movements of the insect during its prey flight. Like a fighter pilot, the dragonfly continuously changes its position to keep its target fly in its sharpest field of vision. Once it has the unfortunate smaller insect in its sights, the dragonfly aligns its body in a targeted manner in the flight path of its prey, reacting to even the fastest evasive manoeuvres of the fly.

The mantis crab is also a skilful predator, and, with its 10,000 ommatidia, it has astonishing visual acuity for an arthropod. Compound eyes also allow their owners to capture light in other wavelength ranges such as ultraviolet or infrared. Insects therefore have a completely different view of the world than we humans do.

PIT AND PINHOLE-CAMERA EYES

Molluscs such as snails are particularly interesting in terms of eyesight because their many different representatives also possess many different eye models, which have evolved over time. The simplest model can be found in limpets: pit eyes, which consist of nothing more than an indentation containing light-sensing cells. As in the pigment cup eyes of flatworms, the light receptors here are shielded by a pigment. This means that the owner of the pit eye can distinguish only the direction and brightness of the light source. The pit eye is the starting point from which further eye types, such as the pinhole-camera eye or the lens eye, may evolve.

In the case of the pinhole-camera eye, the opening of the pit eye gets smaller, so that less light can shine onto the lower layer with the light-sensing cells. This layer can provide an image of a small section of the surroundings. The pinhole-camera eye can be found in representatives of the genus *Nautilus*, part of the cephalopod family. Most examples of these interesting molluscs are already extinct, and we only know them from fossil records. However, a few of them can still be found as 'living fossils' in the western Pacific and some parts of the Indian Ocean. The nautilus has on its head tentacles like those typical of squid, their fellow cephalopods. Its two pinhole-camera eyes give it sufficient vision to live as a predator and look for prey in its surroundings. The nautilus bears the nickname 'pearl boat' because of its mother-of-pearl shell, which it withdraws into when in danger.

The next type of eye is a good example of how a principle that once worked in nature can be found again in a wide range of living creatures – albeit in a modified form. I'm talking about the lens eye: an eye that's equipped with a lens and protective cornea over the light aperture. Such single-lens eyes also occur in invertebrates like some molluscs and show astonishing similarities to the lens eyes found in vertebrates.

Molluscs have a particularly large variety of eye types. The two lens eyes of the Roman snail (*Helix pomatia*) are located on the tips of its tentacles.

BUNDLED VISION THANKS TO LENS EYES

In contrast to the complex eyes of insects, the lens eye has only a single lens, which bundles the light onto the light-sensing cells below. The light initially passes through an opening, the pupil. The pupil is the central 'black hole' in the eye and is surrounded by a ring of muscle – the iris. In its function, the iris corresponds to a camera aperture and can regulate the size of the pupil by tensing and relaxing the muscles.

If we take a closer look at the structure of such lens eyes, significant differences come to light: in an invertebrate animal like the squid, the eye cup is formed from the light-sensing cells together with a layer of skin on the surface of the body. Here, the light-sensing cells are just behind the lens. The lens eyes of vertebrates, on the other hand, have their origin in part of the diencephalon – the middle brain. The light has to first migrate through many layers of cells before it reaches the light-sensing cells called 'rods' and 'cones'. Once there, the incoming light is evaluated. The cones perceive colour, while rods are responsible for perceiving light–dark contrasts. Using complicated nerve connections, the light-sensing cells send the incoming information via the optic nerve on to the visual centre of the brain. This is where the evaluation of information from both eyes takes place. More highly developed animals, in particular, can thereby recognise patterns and determine the direction of movement. Spatial vision is particularly well developed in tree dwellers and predators.

A good friend of mine is also an enthusiastic outdoor biologist and has spent a lot of time in the Indonesian jungle, studying the behaviour of Sulawesian tarsiers. Tarsiers are primates, and their eyes are huge in proportion to their body size, allowing the tarsiers to capture every last bit of moonlight so that they can jump unerringly through the treetops of the nocturnal jungle.

While many animals have muscles to turn their eyes, others have to move their entire head. Owls like the barn owl (*Tyto alba*) can turn their neck 270 degrees – and therefore practically have eyes in the back of their heads. Owls also have an ossified ring around their eyes that's typical to all birds. This ring connects the lens with the layers of skin behind it, like a kind of tube, thereby providing the owl with 2.7 times more light than we humans receive. With these so-called 'telescopic eyes', the barn owl can go hunting for prey even on a moonless night.

When it comes to night vision, cats are also masters. They have built-in residual-light amplifiers in their eyes called 'tapetum lucidum', which means 'bright tapestry'. This 'tapestry' is an extra layer of cells that helps the cat's eyes to capture light. The 'bright tapestry' is also responsible for the demonic reflection of cat's eyes when light hits them. Dogs and horses also have residual-light amplifiers in their eyes.

Left: owls like the barn owl (*Tyto alba*) don't have moving eyes, but they can rotate their heads by 270 degrees. Right: nocturnal predators like the domestic cat (*Felis catus*) have built-in residual-light amplifiers in their eyes called 'tapetum lucidum'. These additional cell layers help to trap the light and thus improve night vision.

Listen and be amazed

We now move from optical information to acoustic information and thus to the question: which receptors does it take to perceive sound? As we have learned, sound is a mechanical oscillation that increases the pressure of the medium, e.g. air or water. On the receiver side, structures are needed to absorb the mechanical energy of the oscillations and start vibrating themselves. Hair and hair-like structures are particularly suited to the job of resonating. They are flexible so they can bend like stalks in the wind with the pressure changes in their environment. Once again, 'resonance' is the magic word here!

MECHANICAL RECEPTORS REACT TO SOUND

With the help of mechanoreceptors on the surface of a cell, living organisms can perceive different types of mechanical or physical forces: elongation and compression forces as well as bending and shear forces. With shear force, we can imagine forces that cause objects or fluids to shift relative to one another. In this way, pressure changes on the surface of a living being are caused not only by acoustic waves, but also by direct touch. Even single-celled organisms such as bacteria can perceive such elongation and pressure forces from their surroundings, as in when they hit an obstacle. Plants and fungi also have receptors for mechanical effects, and plants can even react individually to predators by using these receivers.

Depending on the type of feeding movement, the mechanical pressure that plant eaters exert directly on the surface of the plant's cells differs. These mechanical influences locally change the electrical potential of the receptor, and a real firework of chemical reactions takes place. And that's not all: mechanoreceptors are also found in plant roots and can help our green friends to perceive the movement of water in the soil. But what does this have to do with hearing? Or to put it another way: when I go into the woods and scream, who actually hears me?

THERE IS NOTHING TO HEAR IN PLACES WHERE NO ONE SAYS ANYTHING

Hearing is more than just the resonance of sound through mechanoreceptors found in the ears of animals. Oscillations must be translated into electrical nerve impulses so that they can be passed on to the brain and processed there — so the lack of a brain with an auditory cortex is the reason why unicellular organisms, plants, fungi, and some simply constructed invertebrates can't hear. The same principle at work in vision applies here: sound is heard not in the receptor itself but in the brain once the information from the respective receptors has been received. Perhaps you've noticed that only a few invertebrates play an instrument in our animal orchestra and can therefore actively send acoustic information. In the world of worms and snails, we won't find a sensory hearing organ — there is nothing to hear in places where no one says anything. This doesn't mean that these creatures don't perceive vibrations and thus react to changes in pressure in their environment. Think back to the spiders and insects that are quite capable of producing sounds. Insects in particular are a major exception when it comes to listening because many of them are indeed very musical and produce all kinds of sounds. This would not make sense if they were not also able to receive and process acoustic information.

LOCUSTS LISTEN WITH THEIR LEGS

Arthropods can detect soundwaves with the help of their body hair or with the help of body appendages such as antennae. Depending on their length and rigidity, these simple mechanoreceptors resonate with differing wavelengths. Many butterfly and moth species use their body hair to tune into the same wavelength their predators use to transmit *their* acoustic information. Male mosquitoes even have an acoustic receiver on their antennae that only reacts to the vibrations caused by the flight movements of female mosquitoes!

Crickets and katydids are miles ahead of many other insects when it comes to hearing because of the so-called 'tympanic organ', which is located on their front legs. This is an air chamber covered with a membrane, which works like our eardrum and resonates to any pressure changes in the external medium. A team of researchers from the University of Bristol solved a mystery using the grasshopper species *Copiphora gorgonensis*: the scientists found a structure behind the tympanic organ that bears a striking similarity to the cochlea in the ear of vertebrates. This organ also contains mechanoreceptors in the form of hair cells. Depending on the type of tympanic organ, there may be one, two, or up to 2,000 such sensory cells. Using a laser, the researchers found that not only is the structure of the tympanic organ amazingly similar to a vertebrate's ear, but so too is the way it functions. Successive, lever-like structures guide the sound into the tympanic organ, which is filled with fluid. The hair cells on the membrane resonate as soon as it begins to move with the soundwaves. The movement of the hair cells is then translated into electrical nerve impulses and sent to the insect's brain. The tympanic organ has developed independently in different insects and appears on diverse body parts, like on the wings of many butterflies.

HAMMER, ANVIL, AND STIRRUP — THIS IS HOW A MAMMAL'S EAR WORKS

Most vertebrates have a fluid-filled inner ear with a membrane on which hair-like hearing cells are located. When soundwaves hit the membrane, it begins to vibrate – as do the receptors on top of it. The strength and direction of these vibrations is directly translated into the release of messenger substances on the other end of the sensory cell and then passed on to the brain via other nerve cells. To ensure that the membrane vibrates with even very weak acoustic waves, the incoming soundwaves need to be amplified. There are many stations in the ear for this, consisting of muscles, tiny bones, and membranes such as the famous eardrum.

The basic principle of hearing is almost the same everywhere, but the ears of reptiles (excluding crocodiles), amphibians, and birds are different in their composition and therefore in the number of amplifiers for acoustic information. These creatures don't have an outer ear in the shape of an auricle. Internally, they have one small bone, called an 'ossicle', while mammals are equipped with three ossicles. Some amphibians, like the salamander, don't have an eardrum – instead they just have muscles and skin. Snakes don't have an external ear at all, or an eardrum, and perceive vibrations via their temporomandibular joint, where the jaw connects to the skull.

Let's follow the path of an acoustic signal using the example of a mammalian ear. When we humans hear a sound, the first thing we hear is the pressure changes on the outer part of our ear and thus on the auricle. The auricle works like a funnel: it collects the acoustic vibrations from the environment and concentrates them on a smaller surface. This surface is the eardrum, and it's situated on the border between the outer and middle ear. From the eardrum, the acoustic waves travel across the three ossicles of the middle ear, called the 'hammer', 'anvil', and 'stirrup'. The hammer is connected to the eardrum and transmits the resonance to the anvil. This in turn is connected to the stirrup – incidentally, the smallest bone of a mammal. The stirrup is connected to the entrance port of the fluid-filled cochlea in the inner ear, called the 'oval window'. The path from the eardrum to the oval window has made the soundwaves 15 times stronger in their intensity compared to their surroundings – it was transmitted from a large surface (the eardrum) to a smaller surface (the oval window). Pressure is force per area, and because the same force is now applied to a smaller area, the pressure of the soundwaves on the oval window is greater than that on the eardrum. This pressure is necessary to make the fluid-filled cochlea, and thus the auditory sensory cells lying on it, vibrate. The cochlea is where the translation of mechanical information to electrical nerve impulses takes place, and in turn is connected via nerve cells to the auditory cortex of the brain.

Having ears on either side of the body in vertebrates makes sense: the delay between the arrival of the acoustic waves at each ear, as well as the differences in volume, allows the brain spatial hearing. That's how we can tell where a sound comes from. Spatial hearing is supported by ears that can turn and fold. We have already met the Indonesian tarsiers; along with their huge eyes, these monkeys also have large and especially mobile ears, with which they can hear even the quietest noise. When it gets dark, the little goblin-like creatures open their ears wide and wait for the promising sound of locusts and other treats.

Let's stay in the rainforest and visit harlequin toads in South America. When it comes to hearing and absorbing soundwaves, these amphibians go all out!

Tarsiers living in Sulawesi have rotating and foldable ears, with which they can hear the many different sounds in their jungle habitat.

HEARING WITHOUT EARDRUMS — WHEN SOUND GETS UNDER YOUR SKIN

The elegant stubfoot toad, which lives in the rainforest of Latin America and is also known as the 'harlequin toad', astonished scientists from the Ohio State University. Among harlequin toads there are some species with an eardrum and some species that lack this sound amplifier. Harlequin toads are therefore ideally suited to teach us more about how the hearing ability of amphibians has developed. The scientists played calls of fellow members of the species to toads with and without eardrums. At the same time, they followed the path of the acoustic signal through the toad, which is only 40 millimetres long and weighs just two grams. The scientists measured vibrations on the skin surface at three points. The first was directly above the lungs, the second on the side of the head above the inner ear, and the third halfway between the nostrils and the eyes. All toads showed the strongest vibration in reaction to the soundwaves in the area above the lungs. The skin in the chest area is very thin, so it can easily be made to resonate. Surprisingly, the vibrations were particularly easy to measure when the test frogs heard a call from their own species. This would be as if our entire upper body trembled as soon as someone called us. What was particularly interesting about this study, however, was that further propagation of the soundwaves from the chest area to the ear differed between the species: the toads with eardrums had much stronger vibrations on the side of the head above the inner ear than those toads without eardrums.

Since toads don't have outer ears like mammals, their eardrums lie directly on their head and transmit the acoustic vibrations to the inner ear using hair cells. In those examples of the species without eardrums, it seems that the path of the soundwaves through the air into the lungs and onto the ossicles has proved to be the better option, though it comes at a price. The diversion via the lungs is accompanied by a loss of perception of the highest frequencies – yet this doesn't prevent the little harlequin toads from communicating with each other at frequencies up to 3,780 Hertz. Scientists are still baffled as to how they can do this without eardrums.

WHY FISH HAVE LITTLE STONES IN THEIR EARS

We can't see them, but fish have 'real' ears with which they can hear. What they lack is an outer and middle ear and therefore the sound-transmitting structures that mammals have. The inner ear of a fish is located in the skull behind the ears. As in terrestrial vertebrates, it is also filled with fluid and is the location of the acoustic receptors, the hair cells. But how can a fish perceive changes in the density of the water surrounding it when it's in the water itself and its inner ear is filled with fluid? Shouldn't the soundwaves go straight through it?

Underwater inhabitants solve this problem by getting the soundwaves to transfer from the medium 'water' to the medium 'stone' and sometimes even to the medium 'gas'. Fish have in their ears tiny little limestones known as 'otoliths', which are heavier than the liquid medium surrounding them. When acoustic pressure waves hit the inner ear, the little stones react to these waves with a delayed movement. When an otolith starts moving, it can change the position of the hair cells, which are also located in the inner ear. The mechanical movements of the auditory sensory cells are then passed back to the brain via electrical nerve impulses. This type of sound transmission only works well at low frequencies with few vibrations per second, yet there are a lot of fish that can hear at higher frequencies – what's their secret?

Fish with a bony skeleton have a gas-filled swim bladder, and, thanks to this air cushion, they can easily float through water despite their weight. But the swim bladder can do a lot more. Soundwaves are transmitted from the gas in the swim bladder to its elastic walls, causing these to oscillate. These oscillations continue to spread towards the inner ear, causing the otoliths to move. So the swim bladder amplifies the incoming acoustic waves, thereby functioning as a sort of eardrum. The bigger the swim bladder, the better the hearing ability of the fish. Some cichlid species such as the Indian cichlid (*Etroplus maculatus*) and all species of herring have an additional 'upgrade' that improves their hearing: two bulges on the front end of their swim bladder that are in direct contact with the inner ear. In this way, the Atlantic herring (*Clupea harengus*) can hear surprisingly well within a range of 30 to 5,000 Hertz and can even tell which direction the noise is coming from. As you will see in more detail later, the herring uses very special acoustic signals for communication with fellow members of the species. Carp, catfish, tetras, and knifefish, on the other hand, have tiny ossicles that connect the inner ear and the swim bladder.

THE LATERAL LINE SYSTEM — THE RECEPTION OF ELECTRICAL AND MECHANICAL INFORMATION

Fish and all aquatic amphibians are constantly exposed to pressure waves from near and far. One cause of such pressure waves would be a fish swimming by, or nearby prey, or if the direction of the water flow changes due to an obstacle. With the help of their lateral lines (a sense organ), fish and amphibians can use these pressure waves to collect information about their environment and orientate themselves even in murky water. The lateral lines also help fish in a shoal to maintain the right distance from their neighbour. The mechanoreceptors of these lateral lines are known as 'neuromasts', and are only visible under a microscope. They are hair cells, stabilised by supporting cells and surrounded by a protective gel or sticky mass. The neuromasts are found across a wide area of the skin in fish and amphibians. They also form a system of canals and tubes inside the skin. The neuromasts maintain contact with the water that surrounds them through pores in the skin. If movement in the water causes a pressure wave, this stimulates the sticky mass or jelly surrounding the neuromasts and causes the hair cells to vibrate. This mechanical information is then transmitted to the brain in the form of electrical impulses through the nerve cells. The lateral-line canal is the longest such tube under the skin, and, in some fish, it's clearly visible as a line of fine pores from the gill covers to the base of the tail.

Lateral lines can also pick up nearby sounds. Soundwaves from a distance, on the other hand, don't generate sufficiently strong water movements to cause the neuromasts to perceive them, so the lateral-line system doesn't play an important role in communication when it comes to receiving acoustic information.

The situation is quite different when it comes to the receptor system for electrical and geomagnetic information, which has developed from the lateral-line system. There are canals in the skin of some fish that are filled with an electroconductive substance. These canals are called 'ampullae of Lorenzini' and represent important receptors for communication via electrical signals. We humans know that it's not a good idea to use a hairdryer in the bath, as everyday water is highly conductive. It's exactly this electric conductivity that fish that generate weak electric fields, such as elephantfish or knifefish, use underwater for quick communication. With the help of modified muscle cells or nerve cells on the surface of the skin, those fish living in murky fresh water can generate weak electric fields. Most of them are nocturnal and live at the bottom of their habitat. Eyes would be wasted here, and so, in the course of their development, electrical information was established as the means of communication – for example, to attract a mating partner.

Besides those fish with weak electricity, there are also those with strong electric fields, like the electric eel found in the Amazon. With the ability to generate up to 900 volts, it's obvious why they're called 'electric' eels. As swimming 'stun guns', electric eels – and also electric catfish and electric rays – use electricity to catch prey and ward off enemies. Because they can give such a powerful shock, these fish don't use their electrical ability to communicate, and they don't have receptors for electrical information.

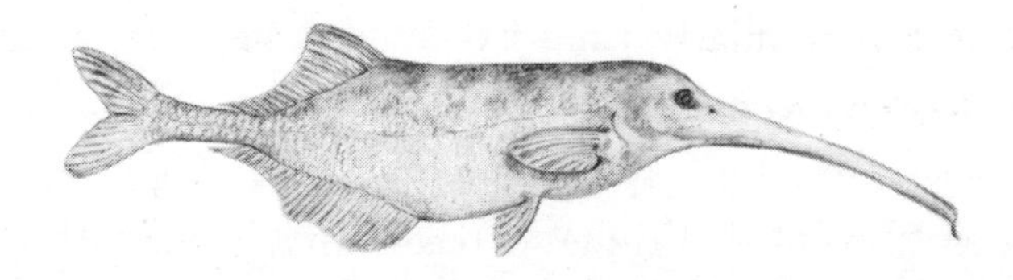

The Upoto elephantfish (*Campylomormyrus numenius*) belongs to the group of fish that generate weak electric fields; it does this by means of modified muscle or nerve cells on the surface of the skin. The elephantfish uses electrical signals to communicate.

Follow your sense of smell

Close your eyes and imagine we are in a forest. You inhale the fresh air with a deep breath. It is summer, and the scent of the forest is particularly strong following a thunderstorm. You can smell leaves, earth, and – this is also part of the forest – the aroma of animal droppings near your feet. The keyword 'smelling' now takes us to the receiver systems for chemical information: chemoreceptors.

THE CHOOSY CHEMORECEPTORS

Among the various receiving systems, chemoreceptors are as old as Methuselah. They fulfil two important functions: smelling and tasting. In their simplest form, they help unicellular organisms to detect chemical substances in their environment. With that help, a bacterium can detect sugar molecules and move towards a delicacy. The chemoreceptors on the cell surface also recognise substances that are toxic to the bacterium – high time for the unicellular organism to move away from the source of danger.

For organisms found on land, chemoreceptors are especially important for receiving information at a distance. We can imagine a fragrance binding to a chemoreceptor the way a key and a lock fit together. Some chemical substances are like universal keys and fit into many chemical receptors. Other substances only fit certain receptor cells, which in turn don't let any passing substance get close. Communication by means of chemical information is therefore a particularly good example of how important fine-tuning between sender and recipient is.

INVERTEBRATES 'SMELL' WITH THEIR ANTENNAE

Invertebrates such as worms, arthropods, and molluscs are particularly dependent on the reception of chemical substances to orientate themselves in their habitat. Their receptors for other information are usually poor to non-existent, and so their communication with other living creatures is dependent on the ability to perceive chemical information. For this purpose, invertebrates have hair-like sensory cells all over their bodies. In insects, crustaceans, and spiders, these are concentrated on prominent body parts such as the antennae or legs. The hair-like protuberances on body appendages increase the surface area so that as many olfactory sensory cells as possible can find space there. This is also the reason why representatives of the cockchafer, such as the field cockchafer (*Melolontha melolontha*), have fan-shaped feelers, on which the olfactory sensory neurons are located. It's no coincidence that insects have two antennae: both have chemical receptors on them, and this allows them to detect the direction a smell comes from.

SMELLING IN VERTEBRATES — A SLIMY AFFAIR

The perception of chemical information is a slimy affair for most vertebrates because it takes place in the olfactory mucous membranes. This 'damp carpet' is located in the upper part of the nasal cavity and is the workplace of the olfactory sensory cells. Depending on the type of animal, things can get tight in the nasal mucosa: in humans, this mucosa is only 2 × 5 square centimetres in size and offers space for about 30 million olfactory sensory cells. In dogs, this area is 100 times larger, and accordingly they have far more olfactory sensory cells than we humans. In contrast to the other receptors, the olfactory sensory cells are replaced and renewed again and again throughout the life span of the animal. The olfactory sensory cells have small hairs on the cell surface, and it's precisely these hairs that the fragrances target.

Let's follow the path of a fragrance using the example of

our own nose: a strong inhalation carries the odours from the air to our olfactory sensory cells. An odorous molecule binds to the corresponding receptor and thereby changes the electrical potential of the receptor. This change is in turn passed on via the activation of axons, which sort, bundle, and send the sense data to the olfactory bulb in the brain. It's only now that you as a living being become aware of what there is to smell.

The lock-and-key principle applies to the olfactory cells in the nasal mucosa of mammals: only one class of odour with a similar structure can bind onto the receptor cells and trigger a reaction. There are many locks and many of each type of lock distributed throughout the mucosa, however, and the varied insertion of keys is merged in the brain to form one olfactory idea. The interconnection of different types enables humans to distinguish up to 10,000 different odours.

The field cockchafer (*Melolontha melolontha*) has fan-shaped sensors on which olfactory sensory cells for the perception of chemical information can be found.

When it comes to smelling in vertebrates, the Danish doctor Ludwig Lewin Jacobson made a name for himself. He discovered an organ later named the 'Jacobson's organ', also known as the 'vomeronasal organ'. It specialises in chemical signals of communication. In most vertebrates, including us humans, the organ has atrophied, and it's completely absent in birds. But for reptiles, the Jacobson's organ plays an important role when it comes to smelling and tasting. The common puff adder (*Bitis arietans*) catches chemical substances in the air with its forked tongue and presses them against the Jacobson's organ in its throat. So for the puff adder, smelling and tasting can be done in one!

PART II

Who exchanges information with whom and why?

3

Unicellular organisms — communication in the smallest spaces

From bacteria in hot sulphur springs and mosses in the cold tundra to fish in the deep, dark sea: life on our earth seems possible practically everywhere. On land, on water, or in the air, it defies even the most inhospitable conditions. So let's take a little stroll through the diversity of life and clarify the question: who exchanges information with whom and why? Let's start at the beginning!

NOISES FROM THE HAY INFUSION

Welcome to my workplace. Please take a seat and make yourself comfortable — I have prepared something. A few days ago, I doused some hay with water from the rain barrel and left the brew at room temperature. Don't worry, I'm not offering you the mixture to drink — even if it sounds like something that people might consume for health reasons. It's actually an experiment that

I'd like to use to show you something.

At first glance, hay seems to be 'dead', because it's the product of dried grass and herbs from a meadow that's been mowed. But hay conceals a secret that's invisible to us: scholars in the 17th century observed that through the addition of water and heat, hay will seemingly come back to life. Grasses and herbs are the habitat of many microorganisms – for example: bacteria. We humans are only able to detect these microorganisms under a microscope and not with the naked eye – hence the 'micro'. If we have 20-20 vision, we can discern a single human hair with a diameter of 100 microns. The dimensions of microorganisms range from 100 microns to a mere single micron. Many of these microorganisms go into stages of hibernation, which allows them to cope with periods of prolonged drought.

By adding water and heat, I've improved the conditions in the hay, and the microorganisms have come back to life like in a zombie movie. Now, they've multiplied to such an extent that it's worth having a look at their microscopic universe. With a pipette, I carefully drip some hay liquid onto a small glass slide and place it under the microscope. I set it at 400 × magnification and can now introduce you to: euglenas, paramecia, didinia, stentors, stylonychia, amoebas, and hay bacteria. Just one drop of hay water is teeming with life, and you have just been introduced to the smallest information transmitter and receiver in the universe: the unicellular organism.

SOLE-TRADER UNICELL: THE SMALLEST TRANSMITTER AND RECEIVER

A unicellular organism is so called because it consists of just one single cell. This small chamber contains everything needed for life, including the ability to take in information. Receptors are positioned on the outer cell membrane to help paramecium and co. find their way around. These receptors react to external pressure. For instance, a hay bacterium can feel if another unicellular organism rams into it.

With the help of their receptors, the unicellular organisms also perceive if I accidentally knock against the slide, thereby creating a seaquake in the drop of hay water. The shock causes movement in the water, which the mechanical receptors on the surface of the cell perceive as pressure differences. The receptors transmit the incoming information to the cell using chemical messengers. A unicellular organism might then react to this information by remaining motionless until the quake is over.

PROKARYOTES AND EUKARYOTES: THE DIFFERENCE LIES IN THE CORE

The two most fundamental versions of unicellular organisms can be found in just a single drop of hay water: the prokaryotic cells and the eukaryotic cells. You can imagine the prokaryotic organism as the simpler and untidier version of a small chamber. Similar to many single-male households, it's sparsely furnished. There's no real 'cupboard', and the few 'possessions' are floating freely throughout the space. Thus, the construction plan of the cell in the form of DNA is also 'unpackaged' and not separated from the rest of the chamber contents. Nevertheless, there's still a system in the prokaryotic organism, and all vital functions are working.

In contrast, the eukaryotic organism is up to ten times larger and definitely more developed when it comes to furnishings — it's an 'upgrade' so to speak. Here, there are isolated work areas within the chamber called 'cell organelles'. One of these new cell components is a real cell nucleus with a shell. The DNA inside is neatly packed and therefore clearly separated from the rest of the cell contents. The other cell components, such as the vacuoles and mitochondria, take over important tasks in metabolism, like determining how much water goes in and out of the cell.

Where does this sudden luxury come from when it comes to cell chambers? There are many indications that at some point large prokaryotes took up smaller prokaryotes and formed a sort of flat share. The cells helped each other in a symbiosis, and, over the course of time, this prokaryote flat share developed into the eukaryotic cell.

The occurrence of prokaryotes and eukaryotes divides nature into three domains:[2] the archaea, the bacteria, and the eukaryota. The archaea and bacteria are made up of simple prokaryotic cells; all other living creatures, like animals and plants, consist of eukaryotic cells. So why are there three domains and not two? After all, there are only two cell types.

PROKARYOTES ARE ARCHAEA AND BACTERIA

Externally, archaea differ from bacteria in some of their characteristics, such as their cell membrane. This likely doesn't mean much to you, but it means a lot to the archaea, which are in some ways closer to eukaryotes than bacteria. They are therefore assigned their own domain. The word 'archaea' is derived from the Greek word *archaios* and means 'ancient things'. Both archaea and bacteria have been around for about 3.5 billion years and are extremely robust – they can even colonise inhospitable areas such as hot and sulphurous volcanic vents in the deep sea.

Small threads known as 'flagella', which are anchored to the cell membrane, allow many prokaryotes to move. A 'motor' is connected to the base of the threads, making them move. Flagella

2 If a species is the smallest stage in the division of life, a domain is the highest level.

can cover the entire surface of the cell or only occur in some areas. When the 'motor' starts, the flagella rotate like a small propeller up to 1,700 times a second, producing enough thrust to propel the cell forward. Alternatively, the cell can engage reverse gear by changing the direction of rotation of the flagella. Flagella enable the sprinter among bacteria, *Candidatus Ovobacter propellens,* to achieve top speeds of one millimetre per second.

Another example of the manifold forms of movement in the microcosm of unicellular organisms is myxobacteria, which live in swarms. A single cell can either produce slime to slide back and forth on or attach itself to a passing neighbour cell – hitchhiking through the bacteria galaxy.

PARAMECIUM AND CO. — UNICELLULAR EUKARYOTIC ORGANISMS

Let's have another look under the microscope: the light from the lamp has heated up the waterdrop and the tiny unicellular organisms are becoming more and more active. A didinium shoots past a euglena, and even an amoeba moves around fast in the hay water with its pseudopod feet. What we humans see with the naked eye as plants, fungi, and animals are living beings made up of many eukaryotic cells. Paramecium and co., on the other hand, are a worldwide band of living things that consist of only one or a few cells of the eukaryotic type. They mainly populate our seas and fresh water and provide an important food source for other living beings.

Part of this band is made up of unicellular algae with the ability to photosynthesise and fungus-like organisms. Some of these microorganisms often take on beautiful forms – above all, the radiolarians, diatoms, and foraminifers. The dead remains of these unicellular organisms are the material from which chalk cliffs once formed.

The shape of unicellular organisms is as varied as their ability to move. Some crawl, some slide, some flow, and some stride. The flagella of the eukaryotic cells differ in structure from those of the prokaryotic cells. In contrast to the flagella found in prokaryotes, the flagella found in eukaryotes are not an appendage but a bulge in the cell membrane and are therefore enclosed by it – a bit like our arms and legs. Amoebas, on the other hand, crawl over the seabed with their pseudopod feet. If their chemical receptors perceive an obstacle on the cell surface, the amoeba adapts to its environment and flows around the obstacle. If a toxic substance binds itself to one of the receptors, a chemical chain reaction in the cell is triggered and the amoeba moves away from the 'source of evil'. I would like to have that much control over my daily life – looks like you've got it made, amoeba!

Chalk cliffs are formed from the calcareous shells of foraminifers, also known as 'forams'. Most species are extinct and only found as fossils. Living representatives of these unicellular organisms of the eukaryotic type are mostly found on the seabed.

FROM UNICELLULAR TO MULTICELLULAR ORGANISMS – THE GREEN ALGAE CHLAMYDOMONAS

On the threshold between unicellular and multicellular, we encounter small living beings with the adorable names *Chlamydomonas*, *Eudorina*, and *Volvox*. It probably all started with the green algae *Chlamydomonas*. This can be found in small bodies of fresh water and, typical of algae, is able to sustain itself via photosynthesis. With its flagella, it is mobile, and an eyespot as a simple light receiver helps it to orientate itself in its surroundings. *Chlamydomonas* cells multiply by simple division, and their daughter cells can survive by themselves. If several *Chlamydomonas* cells remain in contact via cell bridges after they've developed, they form a colony. An outer membrane envelops the colony and holds all the cells together. It seems as if the number 32 is special, because something interesting happens when this number is reached: individual cells get larger, and the eyespot also becomes more prominent. A colony made up of 32 *Chlamydomonas* cells is called *Eudorina* and shows the first signs of a division of labour. The green algae *Volvox* is a particularly beautiful colony of more than 10,000 individual *Chlamydomonas* cells. Surrounded by a gelatinous wall, the flagellated cells arrange themselves in a sphere and communicate with each other by means of cell bridges. Communication between the thousands of cells is vital because, like a ship with oars, the flagella of each cell need to be individually coordinated. If everyone did their own thing, the ship would never reach its destination!

Strictly speaking, *Volvox* is no longer a collection of individual independent cells, because there is already a clear division of labour. In a *Volvox* with 10,000 cells, only a maximum of 16 of them are responsible for reproduction. Once the daughter cells resulting from division are large enough, the whole *Volvox* ball breaks apart and releases the new cells out into the world. These can now form a new sphere, while the remaining cells of the old *Volvox* simply die off. On a small scale, *Volvox* shows how much communication the cells of a multicellular organism require for a living being to function as a whole. But before we continue with the 'big' multicellular organisms like mushrooms, plants, and animals, let's eavesdrop on the communication of unicellular organisms and answer the question: with whom and why do bacteria, paramecia, and amoebas exchange information all day long?

Eating and being eaten

You've just met the green *Chlamydomonas*. If a unicellular organism like the small green algae isn't able to make food for itself through photosynthesis, it has to 'procure' food in the form of other living creatures. The subject of 'eating and being eaten' is therefore one of the most fundamental topics of discussion between living beings of different species. As we saw in the introduction of the book with the example of the cat and the blackbird, predators often eavesdrop on the communication of their prey and use the information they have acquired to their own advantage.

I HUNGRY, YOU FOOD

Bacteria often feed on dead organisms and in this way decompose them to their original individual parts. A new life can then emerge from these parts, and, in this way, bacteria fulfil an enormously important job as nature's garbage collectors. If living beings don't feed on dead organisms or aren't able to produce their own food, they have to find somewhere else to get their daily energy intake. We humans need only stroll to the kitchen, supermarket, or restaurant. There, the food is more or less ready to be eaten, and we don't need to 'convince' our food to come along with us. In the wild, it's quite a different picture: if you were to book a survival course, you'd have to be prepared to search for and catch and kill your own food. You'd be transported back to the time of hunters and gatherers and confronted with questions such as: which mushroom or plant is safe to eat? And how can I obtain food without being eaten on the way? The topic of 'eating and being eaten' is already on the daily to-do list for many microorganisms and requires the sending and receiving of information.

RHIZOBIA ARE NITROGEN-FIXING BACTERIA

Many unicellular organisms make it easier on themselves to obtain food by leaving the soil or water to live in or on other living beings. Bacteria like to make themselves comfortable in skin folds or parts of the body of animals and colonise these in billions, e.g. *Escherichia coli* (better known as *E. coli*) in the human intestine. If a 'give and take' takes place between 'roommate' and 'host' and both are different species, this unequal friendship is called a 'symbiosis'. We came across this term earlier in the chapter, when talking about the peaceful coexistence of two prokaryotic cells. If the two living beings in a symbiosis are as far apart as David and Goliath when it comes to size, the large symbiosis partner is actually called the 'host' (but the small partner is not really called the 'roommate'). Bacteria, for example, support their host's digestion or prevent fungi from seizing power over one or more of its orifices.

One such host is any plant from the legume family, which includes peas, beans, and alfalfa. Like all plants, they need nitrogen for growth and the formation of green leaves. Although the air consists of 78 per cent nitrogen, the unfortunate problem is that plants can't trap the nitrogen that's floating around. But where there's demand, there's always a market! This is why rhizobia pluck nitrogen from the air and convert it into a form that can be used by plants. This ability is a valuable service, and so it's not surprising that many plants enter into an 'agreement' with these nitrogen-binding rhizobia. In return for nitrogen production, the bacteria receive free home delivery of nutrients from the plant. But how do the two parts of this unequal couple find each other?

The roots of legumes send out chemical signals and use this mechanism to attract the rhizobia. Only when the bacterial cells are 'in agreement' on a chemical level with the plant cells can the bacterial cell penetrate the root cell. The transmitter and receiver must send clear information, otherwise there will be no symbiosis. Once the bacterial cell has docked onto the root cell, it can then penetrate it and 'make itself comfortable'. As a reaction to the penetration by the bacteria, the root cells begin to bend and enclose the bacteria. This produces the typical nodules. Interestingly, there seem to be other living beings in the soil that participate in the connection between rhizobia and root cell.

The nematode *Caenorhabditis elegans* is a typical inhabitant of the earth, where it lives off bacteria, among other things. It unerringly finds its favourite food in the vastness of the soil with the help of chemoreceptors on the surface of its body. Bacteria leave a scent trail on their way through the soil, and the nematode simply has to follow this scent and thus directly reach its destination – large accumulations of bacteria in the ground. Yet the chemical scent of bacteria isn't all that attracts the nematode. The leguminous plant called barrel clover (*Medicago truncatula*) also sends out chemical information for which the nematodes have receptors. At first, this seems surprising. Normally, nematodes are unwelcome guests near plants because some eat leaves or transmit pathogens. Furthermore, the nematode's menu plan also includes rhizobia *Sinorhizobium meliloti*, the symbiotic partner of the barrel clover. So why would the clover lure the nematode closer? Its purpose is to use the nematode as a bacteria courier. The 'parcel delivery' takes place in two ways: the first way is directly through the contact of the nematode with the plant root, because there are 'accompanying' bacteria on the surface of the nematode. The second way leads through the worm and takes place via its faeces. The nematode has an active digestion and in an optimal case it can 'deliver' its products, including nodule bacteria, every 45 seconds! Laboratory tests have shown that despite passing through the worm's digestive system, there are still enough living bacteria in the faeces to form a symbiosis with the root of the barrel clover. This little barrel clover is quite clever.

DO BACTERIA HELP LEAF-CUTTING ANTS TO COMMUNICATE?

Bacteria don't only enter into symbioses with plants. Animals also benefit from the microscopic subtenants in many ways. A particularly interesting symbiosis between bacteria and ants literally ran into my path in the dense Mexican jungle. Leaf-cutting ants live there. The name says it all, because the ants cut leaves with their mandibles and feed these to their own cultivated mushrooms inside the ant colony. In turn, the mushrooms serve as food for the ants themselves. But the industrious leaf-cutting ants don't just carry cut leaves around with them: they also carry bacteria on their body surface with whom they live in symbiosis. One function of the bacteria is to keep fungi in check, preventing dangerous diseases from breaking out among the ants. And scientists in São Paulo, Brazil, published findings in 2018 showing that the bacteria provide even greater benefits to the ants than previously assumed: in the case of the leaf-cutting ant of the species *Atta sexdens rubropilosa*, bacteria help them to communicate with fellow members of their species.

Millions of this particular species live together in underground nests. When so many ants live together, they require a particularly good system of communication, so that the logistics work. In the laboratory, the scientists discovered that the ants stored *Serratia marcescens* bacteria in their glands; the scientists managed to isolate the bacteria and discovered that the scent the bacteria produce is similar in its chemical structure to the scent that ants use to mark their pathways or release as an alarm signal for the presence of other ants. Apparently, the bacteria support the chemical communication within the ant nest. The scents in the soil emitted by freely living bacteria are probably how the two symbiosis partners find each other in the first place. The ants recognise the scent of the bacteria as their own and follow it to its origin: the beginning of a lifelong friendship!

THE PARAMECIUM STRIKES BACK

Unicellular organisms such as paramecia are at the top of the menu of many living beings, but they too have developed strategies so as to not get eaten on the spot. The predators of the paramecia betray their 'murderous intentions' quite unconsciously because they emit chemical information. The paramecium has the appropriate receiver on its cell surface and reacts immediately as soon as an enemy scent molecule docks. In response to the presence of the didinium, for example, the paramecium shoots arrows named 'trichocysts'. Trichocysts also occur in other unicellular organisms and are found in their thousands on the organisms' cell surfaces. The weapon serves the paramecium not only for its own defence, but also for catching prey and thus for feeding. Once a trichocyst has been fired, it can't be used a second time. If the paramecium has discovered the approach too late and has already had contact with the attacker, it instead retreats. Other ciliates react to the presence of predators with metamorphosis: they simply change their form in such a way that the predators lose their appetite for them. What was a dainty morsel one minute becomes an indigestible lump the next.

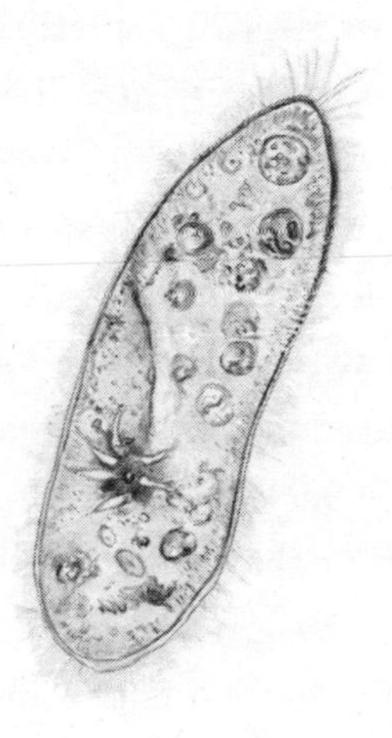

Unicellular paramecia receive information from their surroundings via chemoreceptors on the cell surface.

Says one bacterium to another

Living beings with just one cell are represented in such large numbers on our planet because they practise rapid asexual reproduction. For this type of reproduction, the cell is entirely self-sufficient and doesn't need another cell and therefore no other sex. Yet there are still good reasons to talk about bacteria and sex.

ASEXUAL REPRODUCTION: DOUBLE, WALL, BOB'S YOUR UNCLE!

Under optimal laboratory conditions, *E. coli* can divide every 20 minutes. Let me break down this 20-minute process to ten words and therefore mere seconds' reading time: double the cell construction plan, insert a new wall, done! The result of cell division is two completely identical daughter cells, which have the same blueprint as the parent cell. Plants and some simpler animals like earthworms can also reproduce by dividing themselves. The division can literally crisscross: amoebas, for instance, don't divide on a fixed axis. A cell can also divide multiple times and then decompose into many unicellular organisms. As we have learned, there are cells with and without a cell nucleus. The eukaryotic cells with a nucleus have to divide it too, so that each daughter cell can go out into the world equipped with its own cell nucleus. This division of the cell nucleus is called 'meiosis'. Meiosis enables the growth and constant renewal of an individual cell by division within a multicellular eukaryotic organism. For reproduction of entire multicellular organisms like you and me, you usually need sexual reproduction. But more about this later.

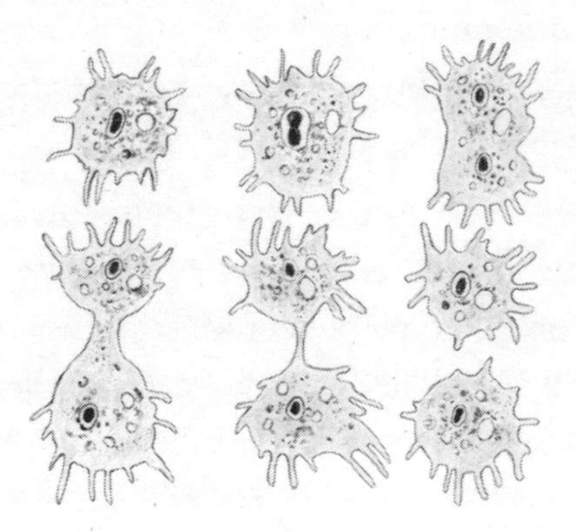

Unicellular organisms like the amoeba reproduce without sex by dividing themselves. Two identical daughter cells develop from one mother cell. (Division from top left to bottom right.)

NEVER CHANGE A WINNING TEAM

Living beings resulting from asexual reproduction don't seem bothered by the fact that they are a carbon copy of their mother. They live according to the motto 'If it ain't broke, don't fix it'. What do I mean by that? Let's use the example of a bacterium that lives in your gut. Here, we have a consistently stable climate at a pleasant 35.5 degrees Celsius. If the mother intestinal bacterium survives in these surroundings with its building plan, the identical daughter cells will survive too – provided that the living conditions don't change. But things can change quickly, like when we take a dose of strong antibiotics. Since all bacteria offspring are identical, it's likely that almost all of them will then die under these changed living conditions. But cell division isn't always accurate, and errors creep in from time to time. One of these errors is mutation. If one letter, one word, or an entire sentence in the cell's building plan changes, the offspring are no longer identical with the mother cell. Such mutations can be lifesaving for the unicellular organism if they prove to be an advantage. However, it can take a few million years until, by chance, the appropriate change occurs.

BACTERIA GET TOGETHER

Bacteria like *E. coli* have developed a mechanism that affords their offspring more individuality – they exchange their DNA blueprint with other *E. coli*! All they require is a small thread-like appendage, the sex pilus. I imagine the whole thing like a science fiction film, when two spacecraft lock together: in the bacterial universe, two 'spacecraft' encounter each other, drive out their sex pilus, and exchange construction plans for their ships. Just finding another bacterium and coordinating the exchange requires certain communication skills. Chemical information plays an important role here, but is that all there is in terms of bacterial communication?

Japanese scientists wanted to find out whether bacteria also react to acoustic information and perhaps even use sounds to communicate. They cultured *Bacillus carboniphilus* in small dishes in a laboratory setting. These quickly formed a colony of many cells, which lived close together in a loose association. The scientists played sounds of different frequencies to the bacteria and were extremely surprised by the reaction. As a reaction to pitches in the range of 6–10 kilohertz, 18–22 kilohertz, and 28–38 kilohertz, the *Bacillus carboniphilus* began to divide, and the colony grew. Even more amazing was the realisation that the bacterial species *Bacillus subtilis* starts sending acoustic information at this pitch, which leads to a crowd of bacteria forming in the lab among the *Bacillus carboniphilus*. Could this be a coincidence? The scientists presume that the bacteria use acoustic signals to stimulate neighbouring cells to divide. This is particularly useful when the conditions under which the microorganisms are currently living become more 'stressful': increased cell division increases the probability of some differences occurring in the building plan of the offspring. For me, that proves once again that everything that is alive sends and receives information and therefore communicates!

4

Multicellular organisms — the language of fungi and plants

Silence in the forest

As to the forest I did go
Today, it was so silent — so —
That when I found
The heart to say
'Here silence truly does abound!'
A whisper only from my lips did stray.

Christian Morgenstern

There is a mixed woodland with beech, oak, and maple trees right outside my study — how about we take a short tour? It's spring, and the small leaves on the beech trees bathe the forest an intense green. The soil is densely covered with moss, and the scent of wild garlic hangs in the air. Contrary to the poem by Christian Morgenstern, a forest is not as silent as it seems at the first 'listening'. It's not only the wind that rustles through the leaves or the rain that drips on blooms. Living beings like plants specifically exchange information with unicellular organisms, fungi, and animals in their environment. This is how flowering plants attract pollinating insects with visual signals. Other plants send chemical signals to enter into a lifelong connection with fungi. This chapter will look at the communication strategies that are used every day by the green beings around us.

TYPICAL PLANT

Over the past millions of years, the algae in water have developed into living beings that are also able to survive outside water: land plants. No longer surrounded by the water that was so essential to them, they developed transport systems such as roots, stems, and leaves. Receptors are on the surface of the plant, and these receive information from the surroundings. Using this information, the plants continually adjust in order to provide themselves with everything they need to survive.

In our forests, trees, mosses, and flowering plants share the typical characteristics of all plant life: sedentariness, photosynthesis, and a rigid cell wall. Sedentariness because land plants don't move from the spot and, apart from mosses (which have only root-like cells), are anchored to the ground through their roots. Photosynthesis because plants produce their own food through solar energy, carbon dioxide, water, and minerals; animals and fungi aren't able to do this and are dependent on other living beings for the absorption of chemical energy. A rigid cell wall because all plant cells have a cell wall in addition to the cell membrane, which gives them stability and prevents excessive water from being released into the environment. As in fungi and animals, plant cells are also eukaryotic and therefore contain a cell nucleus.

A STROLL THROUGH THE PLANT KINGDOM

Not all plants have leaves, roots, and stems. Moss is a comparatively simple version of a land plant and doesn't have roots to anchor it deep in the earth like the 'large' land plants. Mosses, usually only a few centimetres high, grow close to the ground and hold on with their root-like cells called 'rhizoids'. They need a humid environment for their reproduction and absorb all the necessary nutrients from it across their entire surface.

What most mosses lack – a proper transport system for water with nutrients dissolved in it – can be found in seed plants, ferns, and lycopods. With the help of pathways in roots, leaves, and stems, land plants can transport nutrients from the soil up to undreamed of heights. This is why they are known as 'vascular plants'. I was able to see for myself just how high they can grow during a visit to the west coast of Canada. I spent several weeks in the MacMillan Provincial Park on Vancouver Island. There is a large wooded area there called Cathedral Grove. As the name implies, this forest is like a wooden cathedral – it contains Douglas firs over 800 years old that tower up to 75 metres into the blue sky. The circumference of the trunk of such a giant fir can be up to nine metres! Like other firs, spruces, and pines, the Douglas fir belongs to the pine family.

The pine family in turn are gymnosperms, while the flowering plants are angiosperms. We'll take a closer look at the differences a bit later on. Let's stay in the forest for a while longer and address the following question: can you eat that or is it poisonous?

Plants are characterised by sedentariness, photosynthesis, and a rigid cell wall. Above: the branch of an evergreen camphor tree (*Cinnamomum camphora*) with fruit. Camphor trees belong to the laurel family.

FUNGI — NEITHER ANIMAL NOR PLANT

If you order a pizza funghi, then you know what to expect: a pizza with mushrooms. Fungi is indeed the common scientific term for the 'real' mushroom. Originally, naturalists like the Swedish naturalist Carl Linnaeus divided nature into the kingdom of animals, the kingdom of plants, and the kingdom of minerals. For a long time, scholars didn't know what to do about fungi – for a time, they even assigned these living organisms to the kingdom of minerals. Until the late 20th century, it was mainly the sedentary nature of fungi that made naturalists categorise them as plants. Things that didn't move couldn't be animals! Over time, thanks to modern methods of biochemistry and genetics, we humans gained new insights into the diversity of nature. Today, we know that fungi have their own kingdom alongside plants and animals. Like plants, they have solid cell walls, but, because of certain characteristics, they are more closely related to animals.

They contain a substance that until recently was only known to exist in animals – chitin. Chitin is a relatively hard nitrogenous material and gives insects in particular a firm outer armour. Chitin gives more support not only to the physical structure of animals, but also to the fungi. Fungi consist of individual cell cords (known as 'shoestrings'), which make their way through the ground rather like the streets of a city. These cords branch out and can theoretically continue to grow indefinitely, as long as conditions are favourable. These cell cords can take on impressive dimensions. I already mentioned the fungi called *Armillaria ostoyae* at the beginning of the book. It grows underground and occupies thousands of hectares of land in an American national park with its network of cell threads. Hence why it's colloquially known as the 'humungous fungus'.

What we humans call a 'mushroom' and like to eat is, by the way, only the fruit of the mushroom – the actual mushroom is located underground. The fruit body of the 'higher fungi' (the subkingdom Dikarya) results from the interweaving of the mushroom cords to form the most varied shapes: from a bulbous-round mushroom such as the *Bovista* to the hat-shaped umbrella mushroom.

Now that we've had a brief overview of plants and fungi, it's time to look more closely at the communication of these beings. You'll see that the 'talking points' among unicellular organisms, plants, fungi, and animals are basically always the same.

Order a bite!

Plants don't really have a reason to go hunting because, after all, photosynthesis generally allows them to produce enough nutrients to be self-sufficient. However, there are some exceptions who like to put away an 'animal' meal and even depend on the extra portion of protein for their survival. Fungi, on the other hand, have no choice but to feed on other living organisms. Like many unicellular organisms, fungi also often feed off the dead. They populate the forest floor in order to break down leaves, branches, and even entire tree trunks. The common pin mould (*Mucor mucedo*) is particularly fond of bread, while other fungi prefer ripe fruit or animal excrement – there's no accounting for taste. Fungi such as the polypore or the tinder fungus, on the other hand, live as parasites on living organisms. Typical parasites, they are particularly good at looking after themselves and very bad at giving. They use their cell cords to penetrate plant cells and feed on the nutrients produced there without doing any service for the plant in return. Before we really get into this topic, I'd like to first take you to the cinema. There's a film about a voracious plant that I'd like to show you.

BLOODTHIRSTY FOLIAGE IN THE FLORIST'S

I'm a fan of old films, and on the subject of eating and being eaten I immediately think of *Little Shop of Horrors* by Frank Oz. This 1986 film is about a strange plant in an American flower shop. The business has been doing badly for quite some time, and the owner, Mr Mushnik, and his employees Audrey and Seymour have to come up with a plan to avert the threat of insolvency. A particularly exotic plant is meant to give the business a breath of fresh air and attract more customers. The plan works out, and the curious plant in the shop window actually helps Mr Mushnik's flower shop flourish. But their luck doesn't last, and Audrey II – the name they give the plant – soon lets its leaves droop. Seymour takes good care of the customers' favourite, but neither water nor fertiliser is enough to satisfy the plant. There's only one thing that Audrey II likes: blood! A daily dose of red lifeblood allows the plant to blossom again, but its hunger for blood becomes insatiable, and Audrey II takes control of the shop and the florists.

All just science fiction from the American film industry? Not at all! What it shows in an exaggerated form is the daily life of carnivorous plants. These voracious plants are often found in nutrient-poor habitats such as bogs or sandy and stony soil. A meal in the form of unicellular organisms, an insect, or even a small mammal is a useful addition to the menu. Necessity is the mother of invention, and so carnivorous plants are fitted with all sorts of 'hunting tools' in order to finish off their snack. Most predatory animals can kill their prey with muscle power and, if necessary, pursue them, but how does a fixed plant get to its daily portion of meat? What would you do if you couldn't or didn't want to leave the house but still needed food? If, unlike me, you don't live in a tiny village in Brandenburg outside the radius of any delivery service, you could call your favourite takeaway and get the food delivered to your doorstep. All it takes to order are a few bits of information about what you want to eat and where the delivery is going.

THE STORY OF SUNDEW

Carnivorous plants transmit information regarding their food orders directly to their 'meal' while it's still alive – and this happens around the clock! I'd like to use two examples to show you exactly how it works. The first example takes us to a lake near my hometown, which bears the inauspicious name 'Teufelssee' (Devil's Lake). Locals tell scary stories about this place, which is situated in the middle of a forest and is often surrounded by trails of mist. Apparently, the devil meets his witches here at night and they have riotous parties – isn't that a good place for a carnivorous plant? I'll gladly take you to this place, but you must be careful and slowly put one foot in front of the other. The round-leaved sundew (*Drosera rotundifolia*), which is only a few centimetres high, is inconspicuous and can easily be overlooked.

But its innocent name is deceptive, because in the dense moorland it waits for its prey.

The sundew's leaves are covered with many glands that produce a sticky secretion. This secretion gives the plant its name because it starts to glimmer like dew drops when the sun shines on it. The glimmer attracts insects, which settle on the sundew and then get caught there. The secretion is like a liquid glue, and, as soon as the insects lower just one of their six legs on the surface of the plant, it's too late, and they end up as protein-rich insect treats. What happens next is worthy of a horror film: one of the sundew's leaves slowly envelops the prey ... and only opens again when the victim has been completely digested and thus disappeared without a trace.

HURTLING FULL THROTTLE TOWARDS DEATH

Things get pretty dramatic in the midst of the tropical rainforest in Borneo. This is the home of the pitcher plant, and, as its name suggests, this tropical plant looks like a pitcher. The pitcher is formed by a leaf and usually contains a liquid to digest live prey that fall into the pitcher and drown there. The pitcher plant uses many tricks to ensure that its prey actually find their way into the death trap. In the first place, it's attractive to look at, and the edge of the pitcher, in particular, sends visual information that insects are crazy for. The surface of the pitcher rim reflects different wavelength ranges of light than those from the other parts of the leaf and thus clearly stands out. This 'illuminated advertising' on the edge of the pitcher plant makes sense: normally, nectar is formed in the flowers, but, in these plants, the sweet juice sticks to the edge of the leaf like sugar on the rim of a cocktail glass. Some pitcher plants hedge their bets by perfuming the nectar and even the digestive liquid in the pitcher with a scent that's irresistible to insects. This scent acts like a request to the prey to deliver themselves. The nectar also fulfils another task for the pitcher plant.

The cells on the surface of the pitcher rim overlap and, in this way, form small steps that lead towards the inside of the pitcher. When the nectar covers these steps evenly, the edge of the pitcher transforms into a slide. In combination with rainwater, the nectar-covered steps have about the same effect on insect legs as the water film on a road has on our car tyres. The film of water on the surface of the road causes the car tyres to lose their grip on the ground and the car starts to skid. It's this same aquaplaning effect that causes the prey on the surface of the pitcher plant to start spinning. As soon as the pitcher plants have enticed their potential snack with sugary bait and then made it slip, it's pretty much all over. The sweet journey is a one-way street and leads straight into the depths of the pitcher. Once an insect arrives at the bottom, no mountaineering action will help, because the pitcher walls are also much too smooth to provide any traction. Finally, a lid on the pitcher prevents the prey escaping even if it could make its way up the smooth walls.

These tropical carnivorous plants are ideally equipped to digest their prey. With a capacity of up to three litres in the largest species (*Nepenthes rajah*), they hold a deadly cocktail of sour liquid and digestive enzymes ready for their prey. The caught meat in the pitcher plant brings other guests to the prepared table, such as the ant species *Colobopsis schmitzi*. It's able to crawl up the slippery surface of the pitcher plant *Nepenthes bicalcarata* without losing its grip. In addition, *Colobopsis schmitzi* can survive immersion in the digestive liquid of the pitcher plant for up to 30 seconds. For this species of ant, the pitcher is like the land of plenty because often significantly more insects fall into the pitcher than the plant is able to digest. The ants, however, are choosy and only get stuck into large insects from their prey spectrum: assassin bugs, cockroaches, and other ants. They transport these – if necessary, as a team – to the lower edge of the pitcher, which might be up to five centimetres away. There, the ants dissect their prey in peace and quiet and simply let the leftovers fall back into the digestive liquid.

For *Colobopsis schmitzi*, the pitcher plant isn't just a reliable supplier of food. The hollow vines of *Nepenthes bicalcarata* also offer a safe place for building ants' nests. And what does the pitcher plant get from its food-stealing subtenants? The ants know what's appropriate and provide a service to their host in typical symbiotic manner – they tidy and clean! You've probably seen what happens with undigested food – it begins to rot and exudes an evil smell. The regular cleaning actions of the ants help the pitcher plant to keep the rotting processes in check. In experiments, biologists at the University of Cambridge discovered that pitcher plants colonised by *Colobopsis schmitzi* catch twice as much prey as pitcher plants of comparable age without the ant tenants. This is mainly due to the fact that the ants free the edge of the pitcher from the remains of dead insects and other dirt and thus maintain the 'aquaplaning' function.

Let's stay with the pitcher plant for a little longer, because, in my opinion, it offers one of the most fascinating examples of the interaction between plants and animals. The giant *Nepenthes rajah* is a particularly interesting specimen. It's so interesting that a team from the Senckenberg Biodiversity and Climate Research Centre in Frankfurt filmed it for 413.5 hours. The scientists recorded 42 'adult' pitcher plants in the rainforest of Borneo and were amazed when they saw the plants getting regular visits from small mammals. On average, mammals such as the mountain tree shrew and summit rat approached the plants every four hours. What might tree shrews or rats want from a pitcher plant — or put differently, what might the plant want from these mammals with bodies up to 20 centimetres long? In fact, a dead tree shrew was found in one of the examined pitcher plants — could this be a coincidence? The researchers did not rely only on the video footage. In the laboratory, they took a closer look at the scent that the plant gives off and took a sample from the lid of the pitcher plant: they were able to identify more than 44 different scent components! The mix of these scents created a smell that was somewhere between sweet fruits and flowers — a bull's eye in the small mammals' chemoreceptors. What's more, the researchers observed that tree shrews and rats like to use the pitcher plant to relieve themselves. In turn, the animals' excrement and urine attracts flies and mosquitoes. So the pitcher plant's prime target when it comes to prey is probably insects and not necessarily the mountain tree shrew, which is much more difficult to digest — but who knows what secrets the Indonesian rainforest keeps!

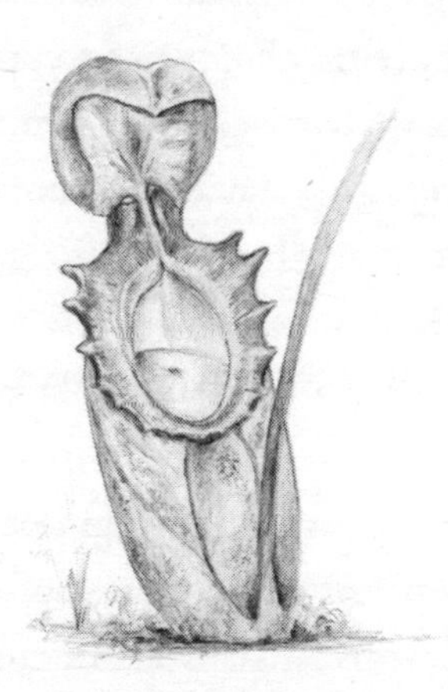

The pitcher plant, which grows in tropical rainforests, is a carnivorous plant. One of the leaves has formed itself into a pitcher, which contains a liquid that can digest insects. The prey slides on the smooth surface of the pitcher plant's rim and can't get out again.

FUNGI, TRAP-SETTERS

We now enter the mysterious depths of the earth with a story I could hardly believe when I first heard about it. Down here, where it is dark and quiet, we come across a very odd communication couple: fungus and nematode. They are connected by the topic of 'food', but I bet you'll be surprised to hear how! Once again, it's the lack of nutrients that makes the sender a carnivore, and – I can tell you this much – the murderer is not the nematode! There are at least 160 species of fungi that are predatory and are out to catch nematodes. But how can a fungus overwhelm a worm?

Think back to the cell cords that the fungus is made of. Some fungi 'craft' these cell cords to create snare traps in the earth that work like a Chinese finger trap. The cell cords initially lie loosely on the soil, but, as soon as an unsuspecting little worm gets caught in them, the thickness of the cord changes. Like a sling, the fungus' cell walls pull together around the wriggling victim. Should we conclude that the worms themselves are to blame for not paying attention to where they are crawling? No, because if there's a fungus like *Arthrobotrys dactyloides* nearby, nematodes such as *Panagrellus redivivus* have no choice but to crawl into the trap. This fungus is one of at least 23 species that send out scents that cause the recipient to change its behaviour, luring the unfortunate worm directly into the trap!

MYCORRHIZA – A FRIENDSHIP BETWEEN FUNGI AND PLANT

The supply of food to plants and fungi does not necessarily have to culminate in drowning and strangulation. Here too there is peaceful contact between two different beings living in symbiosis. Probably the most famous 'love story' of all is between fungus and plant and is called 'mycorrhiza'. Scientists suspect that more than

80 per cent of plants on land enter into an exchange deal with fungi and that these connections have existed for over 120 million years. Depending on how the fungus connects with the plant, we distinguish between ectomycorrhiza and endomycorrhiza.

The fungal partners called 'ectomycorrhiza' use their cell cords to envelop the root cells of their plant partner and partially penetrate the spaces in between the root cells. The network of emerging fungal cells between the root cells is called a 'Hartig net'. Mostly they are higher fungi such as toadstools or porcini mushrooms, which, in the form of mycorrhiza, connect with trees like pines or oaks or eucalypts. The endomycorrhiza, on the other hand, usually occurs between fungi and orchid plants. Here, the fungus' cell cords penetrate as far as the outer roots of the plant and form oval structures. Ideally, the exchange deal consists of the plant releasing some nutrients to the fungus, while the fungus helps its plant partner to retain water and other nutrients from the ground. With its fine cell cords, the fungus thus represents an enlargement of the root surface. Above all, the fungus is a reliable partner for the plant in 'stressful' times because it can give the plant greater tolerance against drought and increases its resistance to pests.

Fungi even help their symbiotic partner to detoxify. They release small molecules into the soil that can bind heavy metals. Talking of 'binding' — how do plant and fungi make this match, and what's the secret to a happy and harmonious lifelong mycorrhizal relationship? You may have already guessed it: communication, communication, and, again, communication! The scaly knight (*Tricholoma vaccinum*) is particularly interesting from a biocommunication perspective because this mycorrhizal fungus speaks exactly the same language as its host plant, forming a symbiosis with trees in mixed and coniferous forests, including those of the common spruce (*Picea abies*). Microbiologists at the University of Jena in Germany discovered that the scaly knight produces the same chemical substance (called 'indole-3-acetic acid') produced by plants to develop their cells. The knight sends out this acid when it wants to 'persuade' its tree partner to grow cells. The more available plant cells, the better the fungus can connect to its symbiosis partner and thus absorb nutrients. But the scaly knight also reacts to the indole-3-acetic acid produced by its tree partner: its cell cords lengthen and form a stronger network. The more the fungus branches out, the closer the connection to its tree partner via the Hartig net. So among other things, it's thanks to this animated dialogue between fungus and plant that the forest has existed as a habitat for the last millions of years.

Plants à la carte

Plants are right at the top of the menu for many living creatures and are therefore at the bottom end of the food chain. While most animals have a choice between fighting, fleeing, and playing dead, sedentary fungi and plants only have the option of fighting! Equipped with an arsenal of spikes, thorns, or nasty toxins, many plants bravely head off into battle and know how to defend themselves against predators of all sizes. And when all else fails, the more communicative among them send out a chemical signal to ask for animal reinforcements.

PLANTS GO INTO BATTLE

It's not that herbivores don't know what they've let themselves in for. The plants put their weapons on public display with a clear message: 'Don't come any closer or you're headed for trouble!' So many plants have sharp-edged leaves, with 'teeth' like a saw blade. Boar thistles, cacti, and nettles keep unwelcome visitors at bay with thorns or bristle hair on their stems and leaves. Reinforced with silicic acid, these hairs are stable like little spears and defend their plant castle against attackers such as snails or caterpillars. Every one of us has probably had some experience with these defence mechanisms: the stinging hairs of a nettle are actually glandular cells and have a very interesting construction that makes them effective weapons. There's a sort of round head on the end of the glandular cell that breaks off easily if touched and gets stuck into the flesh of its attacker. The contents of the head are also emptied here – a mixture of chemical substances that causes an unpleasant burning sensation. Chemical agents are used by many plants, and leading the way are the spring bloomers. Where I live, you need to take particular care when you're in the forest in the months of April and May, when crocuses, lilies-of-the-valley, and snowdrops abound. For humans, the flowers and fruits of lily-of-the-valley can lead to diarrhoea, dizziness, or, in extreme cases, cardiac arrest. The following example of the flowering plant *Arabidopsis* shows how certain plants use their chemical agents to fight off their enemies.

THE EAVESDROPPING ATTACK BY THE *ARABIDOPSIS*

The thale cress (*Arabidopsis thaliana*) is found around the world, a native and naturalised species, and belongs to the cruciferous plants. In particular, the caterpillars of the butterfly species 'small white' or 'cabbage white' (*Pieris rapae*) love to feast on the leaves of the thale cress – but they don't get off lightly! The plant actively fights against the infestation of caterpillars by applying chemical information with one urgent message: 'Get out of here!' In a laboratory, scientists at the University of Missouri tested one possible explanation for how the thale cress might 'know' that the caterpillars are attacking its leaves and that it's high time for defence. Can it really be that the plant recognises the infestation of caterpillars by the feeding noises on its leaves?

First, the scientists recorded the sound of feeding caterpillars on infested plants. They replayed these sounds to a selection of 22 un-infested test plants by means of small loudspeakers. As a control experiment, they equipped further test plants with mini-loudspeakers, but these were mute and didn't emit any eating noises. After two hours of caterpillar audio drama, the experimental plants had reacted with a higher content of chemical repellents in comparison to the control group. The pressure receivers on the leaves are so precise that they can differentiate who or what is causing mechanical vibrations on them. In further experiments, no similar defence reaction came about in reaction to wind vibrations. The sound of the wind through the leaves creates an acoustic pattern completely different from that of the nibbling caterpillars.

But what about noises that are similar to those of the cabbage-white caterpillars? The plants assessed the mating calls of grasshoppers – which, from an acoustic point of view, correspond to the feeding noises of the caterpillars – and deemed them harmless. No similar caterpillar-defence reaction took place.

TOBACCO PLANTS ASKING FOR HELP

Tobacco plants protect themselves from predators such as the insatiable caterpillar with the nerve poison nicotine. Nicotine paralyses the caterpillars or literally spoils their appetite – after all, who wants to bite on an old butt end? In the race between predators and prey, however, there are always ways and means for the predators to circumvent the defence mechanisms of the plant.

The caterpillars of the tobacco hawk moth (*Manduca sexta*) don't mind the nicotine at all – in fact, they love it, and the leaves of the tobacco plant are their favourite food. *Nicotiana attenuata* is the Latin name for a type of tobacco plant that's particularly exciting from a biocommunication perspective. Like the potato or tomato, it belongs to the nightshade family. *Nicotiana attenuata* is also known as coyote tobacco and doesn't put up with the caterpillars of the tobacco hawk moth breaking through its protective barrier – it simply changes its defence strategy!

On the basis of the saliva, the tobacco distinguishes which predator is currently attacking it. As soon as a tobacco hawk caterpillar begins to nibble away at it, chemical substances in the saliva signal to the plant to set biochemical reactions in motion. As a result, the coyote tobacco forms substances that are difficult for the caterpillar to digest. If this isn't enough to cope with the caterpillars, or if they are joined by other predators, then the tobacco sends out chemical signals calling for help. The calls for help are directed at the assassin bug and wasps – who know just what to do. The assassin bug doesn't hesitate to feast on the eggs of the tobacco hawk moth, and also rids the tobacco plant of annoying predators like flea beetles or leaf bugs. The wasps, on the other hand, lay their eggs in the tobacco hawk moth's caterpillars. As soon as the wasp offspring see the light of day, a richly laid table is waiting there for them.

Plants like corn and lima bean also call for animal reinforcements when it comes to fighting against unwelcome guests. If the lima bean (*Phaseolus lunatus*) is infected with the spider-mite species *Tetranychus urticae* and *Tetranychus viennensis*, it will entice predatory mite species by means of chemical signals. The predatory mites *Phytoseiulus persimilis* and *Metaseiulus occidentalis* love to eat spider mites and are only too happy to take up the lima bean's invitation.

But let's go back to the tobacco for a little longer, because there are other insects that give it a hard time — this includes the moth species with the Latin name *Heliothis virescens*. The female moths lay their eggs on the tobacco, and, as soon as offspring, in the form of larvae, hatch, they immediately begin to stuff themselves with tobacco leaves. Of course, the voracious caterpillars don't go unnoticed by the tobacco: not only do they move in a unique way, but they also have saliva that, in its composition, tells the tobacco plant which unsolicited guest it's dealing with. In reaction to so many insatiable caterpillars, the tobacco directly addresses all 'expectant mothers' of *Heliothis virescens*. Laboratory tests uncovered the content of communication between tobacco plant and moth: a tobacco plant already infested with larvae sends out chemical messengers at night that make further adult females of the species remove themselves from an infested tobacco plant and go and lay their eggs on a previously un-infested tobacco plant. The 'wording' sent by the tobacco plant is probably something along the lines of: 'I am a tobacco plant, and caterpillars of your species are already nibbling at me.' The communication from the tobacco plant seems to work — the plants are spared from further attacks by predators, and the female moths find a suitable nursery for their offspring sooner rather than later. The keyword 'offspring' leads us straight onto the next topic, which is also something on the communication to-do list of many plants and fungi.

Sex or no sex

Fungi and many plant species can reproduce both sexually and asexually and therefore have the choice between 'sex or no sex' – but what exactly does that mean? We came across asexual reproduction when we learned about unicellular organisms. A cell doesn't require another cell, and therefore there are no sexes. But what of sexual reproduction? It's reason enough to pull out all the stops, and not only for us humans.

TWO GAMETES MEET

Sexual reproduction has one clear disadvantage compared with asexual reproduction: it takes much longer! A sexually reproducing being must first attain sexual maturity and thus in most cases possess either male or female 'hardware'. The female gametes (ova) are created in the female sexual organs, the male gametes (sperm cells) in the male sexual organs. Only when two different gametes fuse together can this result in a new living being. So that the calculation works out, each gamete contains only half of the DNA construction plan required to build a new cell; this halving of the genetic blueprint of the parent occurs during cell division in the development of the gametes. Sexual reproduction always involves two sexes; however, the fusion of a female egg cell with a male sperm cell is only one possible type of sexual reproduction. Among simpler structures like fungi, we look in vain for 'male' and 'female' and instead find several thousand different sexes. These sexes are called 'mating types', and they form gametes that, in contrast to egg and sperm cells, don't differ externally. And so the fungi gametes are no longer called 'male' and 'female' but become 'plus' and 'minus'. A new fungi offspring comes about when two gametes from different mating types (a plus and a minus cell) merge with one another – so the choice is big!

Scientists are still baffled by the exact steps that nature has taken to get from the several thousand mating types in fungi to just two sexes in plants and animals. In general, the question of the purpose of sexual reproduction has also not been cleared up, because it's not quick for the two genders to get together – as I've said, sex requires time and resources. One of the important reasons why sexual reproduction has nevertheless developed is the wide diversity of descendants. The fusion of the gametes, with half a genetic blueprint each, results in a new cell with a complete blueprint and a new combination of characteristics – for example, the colour of a flower. In this way, offspring (with the exception of identical twins) are inimitable one-offs, unique combinations in the lottery of life. With such great variety among offspring, there's an increased probability of one of them surviving even if conditions change. This possibility of new combinations is the key to the sheer endless diversity of living beings on our planet. Yet not just any blueprint can be merged together with another – a snail cannot reproduce with an elephant, for obvious anatomical reasons. Even if the unequal pair found a way to bring their gametes together, what would come out of two such different blueprints? According to which information should the new living creature be constructed – the blueprint for 'snail' or the blueprint for 'elephant'? So it's only living beings from the same species that can reproduce among themselves: dogs with dogs, cats with cats, humans with humans.

PLANT POLLINATION — HOW POLLEN GETS ON THE STIGMA

Through the fusion of the female egg cell with the male sperm cell, new life also begins in spermatophytes, or 'seed plants'. This is how seed plants got their name – they produce seeds for propagation. The yew and the cedar and the Christmas tree all belong to the gymnosperms, or 'naked-seed plants'. In contrast to angiosperms with flowers, the naked-seed plant's seed is not enclosed in an ovary and is therefore 'naked'. Just think of an apple – here the seeds are well protected by the surrounding flesh.

The flower of an angiosperm has everything that's needed to reproduce in one place: the carpel is the female flower part, and the stamens are the male flower parts. The stamens contain pollen with the male sperm cells. For flowering plants, the sperm cells need to meet the stigma of the carpel, because this is where the ovum is. Plants can pollinate themselves, as when the flower closes and the stamen meets the stigma, but much more important is the dispersion of pollen by wind or insects! If the pollen successfully hits its target, it grows a pollen tube directly to the ovum and releases the sperm cell. The egg and sperm cell fuse and a descendant is produced, which we humans know as 'fruit'. When the fruit is ripe and the seed is ready to be spread, the appearance of the fruit changes. This is how we humans – as well as other creatures – know when that apple, for example, will taste sweet. Colour changes in the UV range also signal to many birds that now is the time to harvest fruit. They eat the fruit, including the seeds; fly from one place to another; and excrete the seeds undamaged through their metabolism somewhere else. A new plant can now grow there.

FLOWERING PLANTS ENTICE WITH REWARDS

The angiosperms, or flowering plants, have found a way to get their gametes to meet with the help of animal 'love messengers'. As this is a matter of life and death, they use the craziest strategies to entice insects for pollination. Flowering plants display plenty of visual information, such as colours, shapes, and surface structures. Using fragrances on top of all this, flowering plants cast a spell on their pollinators – what bee could resist? The shapes and colours of flowers were created in close 'communication' with animals and are often adapted to specific pollinator species. Some flowers are simply monochrome, but most have two colours that contrast strongly with one another. Such contrasts are often found near the sweet reward for the services of a pollinator – nectar! As though it was using a map, the pollinator can orientate itself along the patterns of the flower and easily find what it's looking for. Round and oval dots on the flowers also serve as optical signals and make it easier for insects to orientate themselves on a plant flower.

The wild carrot (*Daucus carota*) is a good example of such striking markings. It has many white flowers, but there's a dark dot in the middle of the flower heads. This dot resembles the outline of a small insect. Experiments have confirmed that flies land more often on the wild carrots that have a black dot than on plants where the dot has been removed. It seems the carrot 'advertises' its popularity among insects and sends the message 'Look over here, good products that have been bought by other customers'. In fact, many plants are 'shrewd businesspeople' and react to their clients' requirements. Some flowering plants signal when they have already been visited by pollinators and have no further need for pollination: they change the colour of their flower or reduce the nectar that's available as a reward for the pollinators. Others shapeshift. Let's briefly fly to Africa to see one of these shapeshifter plants.

DESMODIUM SETIGERUM — THE SHAPESHIFTER AMONG THE PLANTS

The plant with the Latin name *Desmodium setigerum* is a particularly interesting specimen when it comes to pollination. *Desmodium* is the name of a group of plants that are mainly found in Africa. The pollination of the *Desmodium setigerum*'s flowers requires a particular mechanism. The flowers are shaped in such a way that the lower section of the plant offers the pollinating insects a sort of landing platform. Once visitors such as bees have taken a seat, they can search for nectar in peace and quiet. The actual pollination of the flower can only take place, however, when the insect has found the 'secret door' to the inner part of the flower. The movement of the pollinator triggers a tripping mechanism, which instantly opens the stigma and the anthers within the flower. This tripping mechanism can also be found in other examples of the legume family, but, in the case of *Desmodium setigerum*, it has a special task: it serves as a sort of built-in visitor counter.

After an insect has triggered the mechanism, the colour of the flower changes from violet to white or turquoise. At the same time, the upper petal slowly sinks down over the exposed stigma and anthers. The appearance of the flower after pollination therefore changes completely as a sign that the flower has already been pollinated and that the shop is now closed. If you think this reaction by the plant is amazing, hold on to your hat – it gets even better! Normally, a single bee visit is sufficient to pollinate the flower, but, in some cases, re-pollination is necessary because there isn't enough pollen on the stigma. In this case, the already pollinated flower opens up once more to make the stigma visible. The flowers demonstrate this second opportunity to pollinate by putting up an 'open' sign to the outside world: their turquoise colouring becomes more intense and can sometimes even return to the original violet. But why is this *Desmodium* species so set on showing the pollinators which flowers they should fly to and which not? Maybe because the plant is running out of time – the life span of its flowers is only a single day.

Scientist Dara A. Stanley and her team from the National University of Ireland discovered that most *Desmodium setigerum* flowers had already been visited by at least one pollinator by 2.00 pm. By 6.00 pm at the latest, the pollination mechanism had been triggered in almost all the flowers under observation. With its shapeshifting strategy, this plant deliberately guides the approach of the pollinator, thereby apparently ensuring that all its flowers are sufficiently pollinated, even as it's working against the clock. Ingenious!

CHEATING FOR LOVE

Ragworts belong to the orchid family and use a very specific strategy to lure in pollinators: they pretend. Their flowers send out visual and chemical information that's deceptive and thus misleads the receiver. These orchids imitate a female insect by taking on its shape and colour – but that's not all. At the same time, the ragwort emits a chemical substance that's usually emitted by female insects when they're looking for a mate. In the Mediterranean region, the ragwort *Ophrys holoserica* looks like the female of the solitary longhorn bee species *Eucera nigrescens* with its wings open. The plant imitates the female longhorn bee so perfectly that male bees are actually attracted and settle on the lower part of the flower – also known as the lower lip. Once they've landed on the orchid, the males immediately perform their typical mating movements. This is just what the ragwort has been waiting for, because, once again, it has a kind of tripping mechanism that the visitor must trigger for pollination. The male longhorn bee is positioned on the flower under the overhanging pollen packets. As soon as it moves, pollen rains onto its back. The insect's movement not only causes the ragwort's pollen to fall onto the pollinator, but, at the same time, causes pollen already attached to the bee by a different ragwort to be deposited on this one. The male bee therefore moves out of the frying pan into the fire: first it's lured in under false pretences and then it's drenched in pollen! But field experiments have shown that the male longhorn bee is now aware that the flower isn't a female bee. They only fall for an individual orchid's trick once, and never again land on that particular ragwort.

Biologists have given these pretending tactics the name 'mimicry'. We can look at the whole thing as a triangular communication. There's a sender that takes another sender as its role model, and then there's the deceived receiver. In the case of the ragwort, the plant is the sender, the female bee is the role model, and the deceived receiver is the male bee. The ragwort practises so-called 'sexual delusion', but other plants pretend in different ways. Let's take a look at which topics of communication connect the red helleborine, the bluebell, and the leaf-cutter bee.

Ragworts belong to the orchid family and imitate a female bee in the shape and colour of their flower. The male bee is deceived by this and performs mating movements on the lower part of the orchid flower.

THE STORY OF THE RED HELLEBORINE

The red helleborine (*Cephalanthera rubra*), a member of the orchid family, contains colouring agents that reflect visible light in the red spectrum, causing its flowers to appear red. To our human eye, the red helleborine clearly distinguishes itself from the bluebell, whose flowers are usually blue, purple, or white. It is not only their colour that's different: the shape of the flowers also stops us from confusing these two. But if we were to put on a pair of glasses that allows us to see the world from the perspective of a pollinator like the leaf-cutter bee, *Chelostoma rapunculi*, then we would probably struggle to distinguish a bluebell from an orchid. Like most insects, bees can't perceive the red-wavelength range of visible light, and so, from their perspective, the violet bluebell (*Campanula persicifolia*) looks almost identical to the red helleborine. This similarity, which we humans can't see, is no coincidence! The red helleborine imitates the bluebell because its role model has something that the orchid lacks: nectar. In contrast to the ragwort, this orchid is not a 'sexual deluder', but a so-called 'food deluder'. It's a food deluder because it attracts the pollinator with a reward in the form of nectar but doesn't actually keep its sweet promise! The orchid imitates a nectar-rich bluebell so that the pollinators land on it despite the lack of reward. The male leaf-cutter bees are indeed fooled and pollinate the red helleborine without being rewarded for their services.

Once again, this story shows us humans that, in nature, the exchange of information takes place far outside our perception. In biocommunication research, it's always worthwhile looking at things from the perspective of the receiver – who knows how many other deceivers we'll come across in the plant kingdom?

WHY FUNGI CAUSE ANTS TO BURST

Propagation among fungi, which takes place both sexually and asexually, is a much more honest affair. The fungi's cell filaments multiply in the soil through cell division and subsequent sprouting. Fungi can also form small more-or-less-mobile cell packages. These cell packages are also called spores and can survive even in very unfavourable living conditions before new fungi filaments emerge elsewhere. The sexual reproduction of the fungi can also take place by means of spores, but, as is typical for gametes, these only have half a genetic blueprint. For successful reproduction, they must merge with another gamete of the same species. The exact method of sexual reproduction can differ greatly between different groups of fungi. Here's a short anecdote from my childhood as an example.

While walking in the forest, I accidentally stepped on an example of the fungi species called 'puffball'. In response to my misstep, a small cloud shot out of the fungus — much the same way as it would if you stepped on a box of talcum powder. Today I know that the 'mushroom dust' wasn't powder, but a cloud of spores, each containing half a genetic blueprint. The wind or animals (in this case it was me) carry the spores over several kilometres before they sprout in a new and favourable location. Through division, filaments form from the spores, which also only have half a genetic blueprint. So that a new puffball equipped with a stalk and fruit body can grow upwards, it needs the fusion of two fungal threads, which come from spores of different sexes. As you'll recall, fungi are spoilt for choice when it comes to reproduction because they have more than two sexes. But how do the 'lovers' meet? The fungi threads of different sexes simply send out chemical signals, and in this way they can come together.

Other fungi require another living being for reproduction.

This next story takes us into the Brazilian rainforest, where we encounter fungi with the almost unpronounceable Latin name *Ophiocordyceps unilateralis*. They belong to the parasitic fungi and have been nicknamed 'zombie-ant fungi' – and for good reason. In order to reproduce, the fungi take control of a small ant's brain! The spores of the parasitic fungi find their way into the carpenter ant through its food and encounter ideal conditions for germination. Over time, the network of their cell threads interweaves through the entire ant, including its nervous system. Once it arrives at the ant's head, the fungus takes control of its victim and makes it submissive. Carpenter ants (*Camponotus leonardi*) normally live in trees at airy heights of 20 metres. However, when an ant of this species is infested with the zombie fungus, it specifically seeks out leaves that are about 25 centimetres above the ground. How do we know this? In the death grip of the parasitic fungus, the affected carpenter ant bites into the leaves with such ferocity that it leaves imprints. Once the ant has been paralysed by the fungus, the only way this can end is in the big zombie-ant fungal finale: the ant's head bursts because the fungus' fruit body is pushing its way out! Not gruesome enough? Before the grand finale, the fungus steers the carpenter ant with such precision that it settles under suitable foliage to protect itself from rain – which, coincidentally, also happens to be the very place where its fellow carpenter ants have their trails. So when the ant's head explodes, spores from the zombie-ant fungus fall directly out of it and onto their next victim. Sweet dreams!

Lovely neighbours

Trees in the forest

Trees that have long stood close to each other
Soon will crave distance from one another.
So spruce, even oaks, often wish that they might
stealthily, secretly, move and take flight;
but – since to the soil they are firmly bound
with roots long and deep, so people have found,
which force them, like soldiers, to stand and not shake –
their wishes, regretfully, they must forsake.

Heinz Ehrhardt[3]

Concealed from our eyes, plant roots reach far into the soil and there encounter the most diverse neighbours. Biocommunication really gets going down here, particularly with the help of chemical signals. Thale cress, for example, sends out more than 100 different chemical signals, with whose help it communicates with its surroundings. These underground talks aren't always peaceful, but plants get in each other's way above ground too, like when their leaves touch each other in the wind. Even in the plant kingdom, the rule is: love thy neighbour but keep a fence up.

DREAM COUPLE — CHILLI AND BASIL!

Experienced gardeners know the positive and negative effects that neighbouring plants can have on one another. When planting, you have to consider which neighbours suit one another and which ones don't. Onions don't like peas, while fennel feels very comfortable in their company. Why is this so? Plants share the

3 Heinz Erhardt, Noch'n Gedicht © Lappan in der Carlsen Verlag GmbH, Hamburg 2009.

soil with their roots and compete for the available nutrients. Some plants are greedier than others and take up more space or even send out chemical substances that don't agree with their neighbour. The common walnut (*Juglans regia*) is one of these 'nasty neighbours', as its leaves release cinnamic acid, and this in turn hinders growth in other plants.

But there are good neighbours in the plant kingdom too. Basil (*Ocimum basilicum*) is one such neighbour, at least for the chilli plant (*Capsicum annuum*). The basil plant sends out chemical substances that prevent weeds from germinating and growing in its vicinity. It keeps the soil moist and is something like a living mulch supplier for the chilli. Scientists at the University of Western Australia took a closer look at the communication between chilli and basil. They allowed chilli plants to germinate in the presence of basil under various different conditions. In the first attempt, the plants had the opportunity to exchange information both above and below ground via air and soil. In the second attempt, this exchange was prevented and the neighbours were isolated from one another. In both attempts, the chilli seeds germinated better in the presence of basil than without. Why this is and how the chilli 'knew' that a basil plant was nearby without being in contact with it is unclear.

CORN PLANTS PREFER TO STAY ALONE

Plants are wonderfully suited to the exploration of the world of biocommunication. They can be used under controlled conditions and react quickly to changes in their environment. Besides the tobacco plant, corn (*Zea mays*) is also a popular candidate for research into communication between plants both above and below the ground.

Scientists from Uppsala University in Sweden put forward the following question: does contact between the leaves of two adjacent corn plants lead to the release of chemical substances in the soil that may be perceived as information by other members of the species some distance away? In the laboratory, the researchers developed a multi-stage experiment for this purpose. First, they arranged two corn plants so that their leaves had contact, thereby simulating the natural contact between the plants that they would have in a corn field. If there were an underground reaction to this contact, chemical substances for communication would be found in the soil. The researchers then gave the corn plants a choice: would they prefer to grow in the direction of the soil where the two fellow members of the species had touched each other before? Or would they prefer the soil where there had been no previous contact between corn plants? In actual fact, the corn plants extended their roots towards the untouched neighbouring soil. It seems that the contact above ground had released chemical information into the soil that allowed the corn plants to detect the presence of fellow members of the species. We already know from trees that, as soon as they touch neighbouring trees, their crowns do not spread any further.

PLANTS WARN THEIR NEIGHBOURS

Not all plants are as solitary as corn or send out toxic substances to their neighbours like the walnut does. In 1983, scientists observed Sitka willows (*Salix sitchensis*) fighting an infestation of herbivores. Willows closer to a willow already strongly infested were healthier than those that grew further away from an infested specimen. Similar observations were seen in the poplar tree (*Populus × euroamericana*) and the sugar maple (*Acer saccharum*). So do plants actually use the wisdom of the crowd and warn each other if there's imminent danger of an infestation by herbivores? Or to put it differently: are these warning signals being actively sent by a diseased tree or are plants eavesdropping on the chemical reactions of their wounded neighbours? These are exciting questions, but they are difficult to answer. Using the desert mugwort (*Artemisia tridentata*), scientists are trying to learn more about the communication intentions of plants. The desert mugwort also emits chemical substances as soon as it's infested by predators, and, in response, neighbouring plants show a higher content of substances that ward off predators. This is where it gets interesting: the reaction is particularly pronounced between very closely related plants of the desert mugwort, while the reaction did not take place in the presence of other plants that were also infested. So the desert mugwort is therefore able to distinguish between strangers and its own type. Communication using specific signals gives the advantage to its own 'tribe'.

5

Multicellular organisms — communication is animal magic

We're in a forest. Three deer suddenly appear behind a tree about 30 metres away from us. The animals haven't noticed us yet and are looking for food. Their ears turn constantly in all directions and check the surroundings for the presence of danger. One wrong step and the branch under my foot makes the deer prick up their ears. They've discovered us and flee with big jumps into the protection of the dense growth of trees. We've just been able to observe two of the important features of animals: a rapid response to their surroundings through the presence of nerve cells and the ability to move with the help of muscle cells.

WHAT MAKES ANIMALS SO DIFFERENT FROM PLANTS?

Like plants and fungi, animals are also living beings that are made up of a large number of eukaryotic cells. However, animals have several unique characteristics, and one of them is that, in contrast to the cells of plants and fungi, their cells don't have a cell wall, but only a cell membrane as a boundary. Over the course of evolution, the cells of animals have specialised in various tasks, including nerve cells for the transmission of information and muscle cells for movement. In contrast to plants, animals can't live off light and thin air via photosynthesis. They are dependent on other living organisms as a food source and have to find, eat, and digest them. Different animals have different mouthparts enabling them to eat, including teeth, a proboscis, or a radula in molluscs. A sophisticated digestive system, consisting of various organs with all kinds of acidic juices, does the rest.

Predators like the peregrine falcon (*Falco peregrinus*), in particular, need to move quickly in order to keep up with prey such as the nimble mouse. Radar measurements have certified that the peregrine falcon's top speed is 39 metres per second – that's 140 kilometres per hour. Some books even give speeds of up to 360 kilometres per hour when the peregrine falcon dives towards its prey. If you're bombing along the motorway at 120 kilometres per hour, the cheetah (*Acinonyx jubatus*) can easily keep up with you – but only for a few hundred metres, until it gets out of breath. Even creatures that at first glance appear immobile, such as starfish on the seabed, can use the muscles in their feet to cover a distance of a few metres per minute. While falcons, cheetahs, and starfish have as their motto 'I don't do vegetables' and capture other animals for food, pure herbivores feed on leaves, fruits, seeds, or roots. The intake of food is probably the most important reason why animals form information networks with other living beings, including plants. The border between animals and plants is blurred in some areas. As we've seen, carnivorous plants also use digestive juices for the intake of food, while some animals that are able to move don't fancy it much and remain firmly in one place.

A STROLL THROUGH THE ANIMAL KINGDOM — VERTEBRATE OR INVERTEBRATE?

In the animal kingdom, the presence or absence of a skeleton and spine determines whether we're dealing with a vertebrate or an invertebrate. Sponges, coelenterates, worms, molluscs, and arthropods such as everyday insects don't have a skeleton or a backbone and are therefore invertebrates. Their body is often divided into several segments and is proportionally small in comparison to their head. Vertebrates, such as amphibians, reptiles, fish, birds, and mammals, have a skeleton made of bone and/or cartilage that keeps muscles and tendons flexible. This skeleton also includes the aforementioned spine, which supports the torso with two pairs of limbs and, in many vertebrates, also a tail.

Let's begin our short journey through the world of animals with the invertebrate coelenterates such as jellyfish, corals, and sea anemones. They are an exception in terms of movement and are similar to plants in terms of philopatry, meaning they tend to return or remain near a particular site, as they no longer move once they reach maturity. Their simple nerve network of criss-crossing nerve cells is entirely adequate for a sedentary way of life in the sea and enables them to move their tentacles in order to grip food. Invertebrates with a more complex body, such as worms and insects, also have a more complex nervous system. In the individual segments of their bodies, the cell bodies of many nerve cells have been stored together in junctions known as 'ganglia'. At these junctions, the nerve cells can interconnect more easily and enable the coordinated movements that, for instance, the earthworm needs to move through the soil. Especially in arthropods like insects, spiders, or crabs, a lot of nerve cells — which receive and process the incoming information — have gathered in the front segment of the body. This merger of nerve cells in the head was the birthplace of the brain and thus of the central nervous system.

Molluscs colonise a variety of habitats on land and in water and are therefore a prime example of how animals' nervous systems adapt to the requirements of their environment. Firmly attached mussels have only two ganglia, while land snails step up the pace with additional ganglia in their feet, enabling them to glide through your garden in typical snail manner.

Sea slugs known as 'sea hares' (*Aplysia*) are often studied by neurobiologists as they have nerve cells whose cell body is over one millimetre in diameter. Although they have specific ganglia for their external sensory organs (smell, taste, and sight), their internal inner organs (breathing, flight reflex), and movement, their nervous system is simple and therefore easy to study. Neurobiologist Albrecht Vorster from the University of Tübingen in Germany used *Aplysia* to confirm what many students might have already thought: partying the whole night before a big exam is not a good idea! In behavioural experiments, the slugs had to find out how to get to their favourite food, seaweed. The animals succeeded in solving the puzzle a lot better when they were able to sleep the night before. If, on the other hand, the radio was on throughout the night and the slugs constantly had their sleep disturbed, then they did a lot worse in their task the next day.

Incidentally, squid are particularly good at solving puzzles. These animals with mobile tentacles also belong to the molluscs but have made a considerable leap forward with regard to the development of their nervous system. A particularly large number of nerve cells have accumulated in their heads – which is why they are called 'cephalopods', meaning 'head-feet'. And so the eight-legged octopus or the ten-legged squid can not only move through their habitat at lightning speed and catch prey, but also use objects from their environment as tools. Time and again, divers have observed octopuses gathering coconuts and using them to build themselves a protective shield. So cephalopods have an intelligence centre in their brain and are just as intelligent as some vertebrates — which leads us directly on to the next topic.

The central nervous system of vertebrates can clearly be divided into a central and peripheral nervous system. The central part consists of the brain and the spinal cord, while the peripheral nervous system includes all nerve cells that go to or from the brain and spinal cord. These nerve cells transmit information everywhere throughout the body. For example, they conduct electrical signals to muscle cells with the message 'contract'. The brain is responsible for complicated calculations about incoming information as well as for the triggering of appropriate reactions. The spinal cord takes care of the simpler things: this is where reactions to information take place that always follow the same pattern — reflexes. When we touch something hot, we automatically pull our hand back without thinking about it. That's just as well, otherwise we would be exposed to the stimulus 'heat' for far too long before deliberately deciding to take our hand away from the hot object. Such reflexes ensure that the appropriate reaction is carried out in next to no time and, in an emergency, the organism survives. Thus, the spinal cord is in close communication with the brain and in this way coordinates the many reactions that take place in the body. With so much 'computing power', vertebrates are particularly well equipped to explore their environment and to react to the large amounts of information that come in.

A matter of life and death

Predatory animals in particular delve deeply into their bag of tricks in order to creep up on or attract their dinner. They even go so far as to eavesdrop on the communication of their prey in order to learn their language and use it for their own purposes. If there are hungry mouths waiting to be filled at home, then any means of getting food seems justified. Even intentionally sending false information is not unusual. And so the encounter between predator and prey ends with anything but a nice chat over the garden fence – this is a matter of life and death!

SPIDERS ARE THE MASTER PREDATORS

Arthropods like insects, crustaceans, and spiders are not only welcome food items for other creatures – many of them do a great job of catching food themselves and are excellent predators. From sticky webbing to venomous fangs, spiders have an entire arsenal of hunting weapons. They perceive the vibrations of their prey and use such mechanical information to approach their victim without being noticed. As soon as other arthropods set one foot on the artwork of sticky threads of a spider's web, it is too late and there is no escape. You might think that it's the prey's own fault if it doesn't watch where it steps, but spiders don't rely on their prey being careless. Some spiders' silk reflects light in the ultraviolet range and attracts insects in a targeted way.

Some Australian orb-weaver spiders, also known as 'bolas spiders' or 'fishing spiders', use a different strategy: instead of an entire spider's web, they use a sticky 'capture blob' of silk on the end of a line. This is known as a 'bola'.[4] The bola is also perfumed

4 A bola is a throwing weapon consisting of interconnected cords with weights, such as balls, attached to the ends.

with a scent that's astonishingly similar to the sex pheromone of female moths. The spider swings its hunting weapon through the air in the hope that a moth will be unable to resist and will get caught on its sticky snare. This scent spoof is called 'attack mimicry': the predator or parasite sends out visual, acoustic, or olfactory information that ensures its prey can't react in any other way than to surrender to its fate.

FIREFLIES SEND FALSE LIGHT SIGNALS

I already raved about the glowing larvae of the fungus gnats in the Waitomo caves in New Zealand. The species that live there, *Arachnocampa luminosa* – to the Maori, they are known as *titwai* – use sticky threads to catch their prey, in the same way as the fishing spider. The larvae build themselves a tubular-shaped nest of silk threads that hang from the ceiling of the cave. Sticky silk threads are attached to the tube at five-millimetre intervals; these hang down like fishing lines to a length of around 50 centimetres. To ensure prey does actually get caught in their nets, the larvae use their ability to generate light through bioluminescence, attracting insects such as moths. All sorts of delicacies get caught in their fishing lines – ants, millipedes, small snails. The larvae know how important it is to keep your hunting equipment in good condition: after the meal, they remove all food residue from the silk threads so that these can reach full adhesive strength again.

Let's follow the trail from the glowing larvae of the fungus gnats in the caves in New Zealand to the fireflies on the Japanese island of Hokkaido. A few years ago, I attended a conference in Sapporo, the capital of Hokkaido. The program included a visit to the nature centre on the outskirts of the city, and there were fireflies there to marvel at. The highlight of the trip was the short night walk to the nearby park of the nature centre to observe the live communication of the animals known variously as 'fireflies', 'lightning bugs', or 'glow-worm beetles'. Stumbling down the dark steps to the river was quite an adventure – but it was worth it!

Like small lanterns, thousands and thousands of the shining insects whirred through the night and lit up the moonless sky. Each firefly species has its very own light signal, which allows the male and female to find each other. However, right now we are not looking at the communication topic 'My place or yours?' but at the momentous lie that ends in death for many male fireflies. There is a firefly species in North America whose transmitted visual signal signifies anything but a lover's tryst. The female of the species *Photuris versicolor* is able to send the visual mating signals of four other firefly species alongside her own. By using these signals that are not her own, the female entices the males of other firefly species. In at least one out of ten attempts, the female is successful, and the foreign male takes up her invitation in joyful anticipation of meeting a female from his own species to mate with. By the time the male has recognised the scam, it's too late: the female from the foreign firefly species makes short work of it and greedily devours him. Well, it's one way to enjoy a candlelit dinner!

THE COMB-TOOTH BLENNY: NOT SO MUCH CLEANING AS POLISHING OFF

We now move from arthropods to fish and therefore the largest group among the vertebrates. They colonise every body of water on our earth, no matter how small, regardless of whether it is fresh or salt, tropical or arctic. Fish come in many different forms and colours, and their feeding habits are also very diverse. Let's visit the comb-tooth blenny, *Aspidontus taeniatus*, on the Maldives. This little fish is just 15 centimetres long but lives up to its name, with comb-like teeth lining their jaws. It imitates the appearance and behaviour of another fish — the blue-streak cleaner wrasse, *Labroides dimidiatus*. Wrasses earn their living in an honest way by removing dead skin, parasites, and food residue from other fish. Their 'clients' recognise the cleaner fish by its very specific way of swimming. This visual information is so striking that biologists called it the 'cleaner dance'. The comb-tooth blenny imitates the wrasse so convincingly that it can pass as a wrasse and approach other fish without endangering itself. Once the comb-tooth blenny has gained the trust of the client and is close enough, it starts removing whole shreds of skin from its prey with those sharp teeth. Instead of 'cleaning', it polishes off its prey, and the only one to benefit from this is the comb-tooth blenny!

WHEN FISH GO FISHING

As the daughter of a passionate fisherman, I know the patience that's required to catch food by going fishing. And you don't just need patience but also the appropriate equipment — from a rod and bait to the fisherman's camouflage clothing. Of course, the place and the time also determine the best catch, and who would know more about this than a fish itself? There are a multitude of fish such as the warty frogfish, leftvent, and black seadevil who have themselves become 'fishermen'. The warty frogfish (or warty anglerfish, *Antennarius maculatus*) and the deep-sea anglerfish belong to the order of Lophiiformes. These are bony fish that almost all live in the sea, whose pectoral fins are positioned at such an angle to their body that they look as if they are wearing small sleeves. By cleverly coordinating the pectoral fins with their ventral fins, these fish can even manage a brisk gallop across the seabed — though in relation to all other galloping animals, they are definitely the slowest.

Almost all anglerfish live in the sea, but there are big differences with regard to depth: warty frogfish are often found in coral reefs and live in shallow water, while deep-sea anglerfish don't cast their line at anything less than 300 metres. A shallow water channel provides warty frogfish with light, and they can therefore use different bait than their deep-sea fishing colleagues. The foremost spine of the dorsal fin is located on the fish's head and this skin appendage hangs directly in front of the fish's mouth, serving as a 'fishing rod'. Depending on the type of anglerfish species, they offer their prey different bait – from worm to shrimp and even other fish. Yet the best bait is of no use if the fisherman is recognised as such! So in shallow waters, warty frogfish have adapted their shape and colour to their surroundings and are well camouflaged – both to make a big catch and to stay undiscovered by predators. Lured by the treat, the guileless prey swims directly in front of the mouth of the warty frogfish and, when the right moment arrives, is consumed in one lightning-fast bite. The issue of camouflage does not come up in the deep sea, but the question remains as to which bait the deep-sea anglerfish should use. Once again, it is bioluminescence that is used to attract prey. Well then, good fishing!

USING ULTRASOUND TO SEARCH FOR FOOD

Another strategy to get to food is to actively seek out prey – for example, with the help of soundwaves. Many mammals such as dolphins, whales, and bats send out calls in the ultrasonic range to track their prey. Quick reminder: at ultrasound level, the soundwaves oscillate faster than 20,000 Hertz and therefore lie outside the range of human hearing. If the emitted soundwaves meet the prey, its body reflects these soundwaves, and they are returned to sender. The predator can draw a lot of information from this unintentional 'reply' from the prey, e.g. how far away the prey is. Bats first send out echolocation sounds in large intervals. Once the animals have located their prey, they call out in ever shorter intervals. Based on the volume of the returning ultrasonic waves, the bats can determine the size of the prospective prey. However, this type of foraging has two disadvantages: the range is not very large, and the ultrasonic calls only allow the transmitter to reach a narrow area. Prey animals such as moths perceive the bats' calls with their antennae and evade them by simply letting themselves fall to the ground. The barbastelle bat (*Barbastella barbastellus*), with its characteristic pug-shaped nose, is aware of its prey's excellent hearing and, shortly before it is about to attack, sends out a signal that is so quiet that the moths don't hear the bat coming. As the following story shows, using quiet tones when hunting is just one strategy for predators to approach their prey unnoticed.

WHY SILENCE SOMETIMES REALLY IS GOLDEN

In the north-eastern Pacific off the coast of Canada and the USA, scientists discovered two different types of killer whale (*Orcinus orca*). One type of killer whale is described as a 'resident'. It lives in consistent groups and has an absolute preference for salmon. The other type has less of a preference for salmon and prefers warm-blooded prey such as seals, sea lions, and dolphins. This type was named 'transient'. Both types use short successive clicks in the ultrasonic range for orientation and catching prey as well as pulsating calls to communicate among their species. In underwater experiments, scientists at the University of Victoria on the west coast of Canada discovered that the transient killer whales, with their predilection for seals, are less communicative than their fellow fish-lovers, the residents. The transients send out a comparable number of pulsating calls as the residents only when fellow members of the species can be seen on the surface of the water — and now it gets interesting — having successfully captured prey. The proverb 'Speech is silver, silence is golden' seems to apply to the transients' hunting pattern, because the transients' prey, such as dolphins and sea lions, can hear the pulsating calls of the killer whales from several kilometres away, and it's only if the whales don't give off any calls that they have a chance to successfully capture prey. Once the hunt has been successful, the transients once more use acoustic signals to communicate. The residents, on the other hand, prefer salmon as food, and, due to the fish's limited capacity to hear, the salmon can't hear the whales' calls in the first place.

Killer whales (*Orcinus orca*) use short successive clicks at an ultrasonic level for orientation and for catching prey, as well as whistling tones and pulsating calls to communicate. The proportion of black and white segments on the body is unique to each animal and allows for individual differentiation. Shown here is a male animal (above) and female animal (below).

DOLPHINS CATCH FISH TOGETHER — AND CALL EACH OTHER BY NAME

Dolphins can not only hear the calls of killer whales, but also transmit acoustic information between themselves for foraging and other communication purposes. Their prey are large shoals of fish, which they find with the help of ultrasound. Dolphins, such as the bottle-nose dolphin (*Tursiops truncatus*), have many hunting techniques – one of them is hunting together with humans. In the city of Laguna in Brazil, a group of 55 bottle-nose dolphins repeatedly drive fish towards the beach and therefore into the arms of the local fishermen. These fishermen eagerly await the dolphins and their gifts. The fishermen stand motionless in a line, up to their hips in water, with the nets ready. Using head movements and tail strokes, the dolphins communicate with the fishermen and make them understand where and when they should cast their nets. In return for their help, the fishermen give the dolphins the small fish that are able to free themselves from the net. Dolphins as young as four months old take part in this unique hunting strategy and have learned to communicate with people.

But how do the dolphins find each other when they set out to hunt – do they call each other by their name? A team of researchers from the University of St Andrew's in Scotland examined exactly this question and found out that dolphins really do give each other names and use them. Dolphins send out high-pitched click and whistle tones that echo up to a distance of 20 kilometres. Each animal has its own unique tone!

PARASITES — GOOD AT TAKING, BAD AT GIVING

Let's now turn to a completely different story, one that leads us into the world of parasites and their hosts. Parasites are living beings that live in or on another living being – the host. The host is usually much larger than the parasite and serves as a habitat and food source. The parasite makes the most of its host, taking its blood or feeding on its organs. Usually, the host doesn't die from the presence of the parasite, but it depends on the 'dose'. There are endless examples of interaction between host and parasite, but one story particularly impressed me during my studies. It concerns the small liver fluke *Dicrocoelium dendriticum*.

A SMALL LIVER FLUKE ON THE MOVE

There once was a small liver fluke called *Dicrocoelium dendriticum*. The liver fluke felt particularly at home in the bile ducts of sheep, goats, hares, rabbits, and dogs. If the little leech lacks for nothing and is completely satisfied, it diligently produces eggs. Carried by the bile, the eggs leave their warm and safe home inside their host during its next bowel evacuation – in the case of sheep, up to 5,000 eggs can be found in just one gram of faeces. The eggs have big plans and want to discover the world! Naturally, mother fluke has done a good job in preparing the little ones: safely packed, the eggs are protected against the cold, harsh world and can even survive the winter. So the eggs lie around and wait and wait and wait. For what? They are waiting for the free ride called 'snail' – more precisely, the air-breathing land snail. The snails use their radula (toothy tongue) to rasp off the subsoil as they search for food. If the snail gets unlucky, it gets hold of a blade of grass with the liver fluke eggs on it. This is how the eggs get into the snail's body. The eggs contain a secret called 'miracidia'. These are the larvae of the small liver fluke – a sort of pubescent preliminary stage to the adult animal. The miracidia slip out of the eggs in the snail's intestine and form a skin around themselves. In this way, they are well protected from the influences of their host. With this cloak of cells, the miracidia transform into first-order sporocysts (elongated sacs). These then divide and create second-order, or daughter, sporocysts. After a further division, our liver flukes take on another identity and are now called 'cercaria'. The cercaria spend the next three to four months dividing happily inside the snail, and they all live happily ... but wait a moment, the story is not yet finished!

We haven't reached the mature, adult creatures yet, as the cercaria are also larvae and thus precursors. As soon as they are fully grown, they get wanderlust and set off on a journey. Their goal is the snail's respiratory cavity, with a stopover in the pancreas. Like small mountaineers, the cercaria use their hooks to climb up the breathing cavity of the unsuspecting land snail. Once they reach the top, the snail detects their presence and produces mucus to rid itself of these unpleasant guests. A two-millimetre slime ball now contains around 4,000 cercaria ready for departure. The cercaria leave the snail together with the slime ball and then have to 'hurry' because they have an expiration date in the outside world of just a few days: the slime balls lie around lazily in the grass and – you guessed it – wait for their next host. Such a slime ball is a welcome snack for ants, but this fast food comes at a price ... once the ant has wolfed down the slime trojan horse, it's too late: the small liver fluke's cercaria have reached their destination! They settle down inside the ant and make themselves comfortable and, during the next two months, develop into the next stage, called 'metacercaria'.

However, some cercaria just can't wait, and they go exploring inside the ant. They move from the stomach towards the head in the direction of the part of the ant's nervous system called the 'sub-oesophageal ganglion'. This ganglion steers the ant's mouth parts. You can probably guess what happens next. A single small-liver-fluke larva can take complete control of the mouth to make the ant a submissive slave. Remotely controlled by the cercaria, the ant's behaviour changes. Normally, the ants withdraw to their nest as soon as the temperature drops below 15 degrees Celsius in the evening. But one of the larvae doesn't even want to think about going to bed. Instead, it drives the ant to climb up the next nice blade of grass and bite into the tip. The ant has to obey these instructions and spends the entire night on the tip of the blade of grass. As soon as the temperature rises the next morning, the ant's convulsive grip is loosened, and it continues with its daily routine as if nothing had happened.

The question is whether the ant actually makes it to the next morning. Clinging to the blade of grass like this, it only needs the next sheep that comes along to take a big bite of grass, and the ant is history. Together with the ant, the larvae have then made their way back into a sheep (or goat or hare or rabbit or dog). At this point, the cycle has come full circle: the metacercaria in the ant find their way into the bile ducts of the final host and grow into small liver flukes. As soon as these produce eggs, the story starts all over again! Six months have gone by from the first egg at the beginning of the story up to this moment.

The story of the small parasite is an unexceptional example of a sequence of communication events so perfectly timed that it almost seems like witchcraft. So many conditions and circumstances are necessary for the survival of the small liver fluke, and yet – it works!

Coming, ready or not

Where's your focus right now? Still with this book or already at tomorrow's meeting, tonight's dinner, or the weekend? If this sounds like you, it may be a comfort to know that being in constant dialogue with ourselves is absolutely human. As civilised beings, we walk through the streets, lost in thought, and run straight into lampposts and other people. In the wild, if you stop paying attention, you could end up as someone else's dinner. Let's change perspective and slide down the food chain a little to those who get eaten. Welcome to a world where danger looms at every corner and where the inhabitants never know whether they will survive the next moment. Welcome to communication from the perspective of prey animals!

THE TRUTH ABOUT SPONGEBOB SQUAREPANTS

There is an American cartoon series called *SpongeBob SquarePants*, whose hero is a sponge living in the sea. Sponges belong to the family of invertebrates and have no head. The hero of the series wears a collar and tie and lives in a pineapple. The collar and tie are actually not so far-fetched, because real sponges have collar cells that they use to filter out algae from the water. But while the hero of the series lives in a pineapple, the sponges in the wild prefer to make themselves comfortable on a coral. Sponges also don't jump around, but have rather a sedentary lifestyle. Like plants and fungi, they can't run away from their enemies, but have to face them bravely. As harmless as the little sponges appear at first sight, they are anything but defencelessly exposed to their predators. First of all, many of them are covered all over with spines made of lime carbonate or silica, which are part of their body armour. These spines are indigestible to predators — who likes to bite on toothpicks, right? The larger the sponge spines are, the better they keep away large enemies, like the blue-headed wrasse, *Thalassoma bifasciatum*. The smaller sponges also use chemical weapons and send out poisonous substances to keep their enemies at bay. And that's the real truth about our kids' little yellow friend.

DON'T LICK SNAILS

Along with snails, the mollusc family includes all types of mussels as well as octopus and squid. Most molluscs carry a protective shell around with them and live in water. 'Don't lick' is generally good advice for creatures that target molluscs without a 'house'. The slime on the surface of many slugs is doubly effective: it provides a physical shield and is often combined with 'chemical-warfare agents'. Despite their defence mechanisms, most molluscs are nevertheless popular prey for many predators because they're not very speedy.

However, using the example of the pilgrim mussel (*Pecten jacobaeus*), scientists have shown that snails can also flee. The natural enemies of the pilgrim mussel are predatory starfish such as the common starfish (*Asterias rubens*). If the biologists added extracts of predatory starfish to the water, the mussels immediately reacted as if they were really under attack: they closed their shells or set off with big leaps and quick swimming movements. Extracts from starfish harmless to mussels did not cause a similar panic.

Many snails use a completely different survival tactic in the presence of their archenemy, the crab. Crabs use the strength of their claws to grasp the soft interior of mussels and snails – so the molluscs need to be particularly wary of them. But a crab that's on the hunt reveals its presence by its smell. The snails perceive their enemy from a great distance and prepare for the arrival of the predator by first of all stopping feeding. In behavioural experiments that lasted several months, the common mussel (*Mytilus edulis*) and the flat periwinkle (*Littorina obtusata*) adapted to the scent of predatory crabs by developing a thicker shell. So how can predators even set out to hunt if the excretions of their metabolic products announce their arrival? Biologists assume that adaptations are also taking place on the part of the hunters and that they're finding ways to reduce their emissions and thus give away less treacherous information to their prey.

YOU CAN'T SCARE A SEA HARE

Let's stay with the defensive mechanisms of molluscs for a little bit longer. I've already noted that sea hares are neurobiologists' favourite study objects. These sea slugs owe their name to two feeler-like features on their heads – the rhinophores – which look like rabbit ears that are sticking up. The rhinophores allow the sea hares to perceive the movement of the water and are also special receiving stations for chemical substances that the slugs use to communicate with each other. The sea hares are therefore also particularly important study subjects from the point of view of behavioural biologists. Some examples of the species, such as the Californian sea hare (*Aplysia californica*), keep unpleasant guests away by using a fogging tactic. If an attacker approaches them, the sea hares send off a charge of violet ink. This intense colour cloud serves to muddle the attacker's senses, and signals to fellow slugs that danger is imminent! The snail obtains the raw materials for the violet ink from its diet of red algae. This is stored in the sea hare's skin and makes it an unappetising snack for predators such as fish or birds – that's why it's not easy to scare a sea hare.

CHEMICAL WARFARE AGENTS IN THE INSECT WORLD

As the following trip into the countryside shows, it's not only snails that deploy chemical information as a defence mechanism. Imagine a lovely summer's day, the picnic blanket laid out, the delicious food unpacked, and everything seems peaceful. Then you notice: one ant, two ants, too many ants! You have set up camp near an ant nest and now you are public enemy number one. The ants are extremely unhappy with your presence, and they tell you this by deploying a stinger full of formic acid. This has an immediate effect, and you pack your things together faster than you can say, 'Ants in your pants.'

Formic acid is also known as 'methanoic acid' and is not only used by these small insects to ward off big enemies. It's also used as a defence mechanism by stinging nettles and makes our skin burn after it has come into contact with the small hairs on the plant. But humans also appreciate formic acid: we use it in the production of alcohol and to disinfect wine and beer barrels. Formic acid also finds its way into our juices and gingerbread, under the abbreviation 'E236', where it acts as a preservative. In the past, formic acid was actually extracted directly from ants, as described impressively in a historical text by the doctor Christoph Girtanner from 1795: 'The formic acid is obtained by distillation from the ants (*Formica rufa*). The ants are distilled over a gentle fire and the result is formic acid. It accounts for about half the weight of the ants. Or you can wash the ants in cold water, and then place them on a cloth and pour boiling water over them. If you press down gently on the ants, then the acid intensifies. In order to purify the acid, it is necessary to repeatedly distil it, and in order to concentrate it, you need to freeze it. Or even better: you collect ants, squash them without water, and then distil the acid from this.' Luckily for ants, times have changed!

BEETLES LADEN WITH EXPLOSIVES

The following story of the bombardier beetle, from the family of ground beetles, is another example of how living organisms send chemical repellents in the fight for survival. If the bombardier beetles are attacked by a predator such as an ant, they don't waste much time in declaring war. The defensive ground beetle shoots poisonous gases right into the 'face' of its attacker in order to put the predator to flight. The bombardier beetle uses a similar technique to that used by the Germans in World War II for the propulsion of the infamous flying bomb, the V-1. The bombardier beetles need to proceed with extreme caution when they handle explosive materials! They have everything they require to build a bomb in their abdomen: a gland, a collection bladder, and an explosion chamber. So that the beetle doesn't blow itself up, it needs to detonate the bomb at the right time. Shortly before the attack, the beetle sends a reaction starter into the collection bladder, which contains the explosive chemicals. Once the reaction has started, a lot of energy in the form of heat and high pressure is set free. This combination is doubly effective: the high pressure hurls the mixture, at a heat of 100 degrees, onto the attacker with a loud 'popping' sound. The beetle can move its abdomen so flexibly that it doesn't need to turn around when it fires. Bombardier beetles only use part of their powder for one shot. This is the reason why the species *Stenaptinus insignis*, which is found in Kenya, is able to reload particularly quickly and fire off up to 500 'bombs' per second.

Stink bugs are generally also not amused by being threatened. Hawthorn shield bugs (*Acanthosoma haemorrhoidale*) are non-toxic and therefore theoretically a healthy treat. But they keep their attackers at a distance by sending off a foul-smelling odour. This odour is so effective that it even manages to keep birds away from the much smaller bug.

TOAD YOGA

Many animals use visual information to ward off their enemies – even we humans understand this language: would you casually walk up to a wolf that's baring its teeth, a hissing cat, or a bear that's standing upright? One particularly interesting technique of threatening behaviour is shown by toads, whereby they combine both colour and movement. The yellow-bellied toad (*Bombina variegata*) and the fire-bellied toad (*Bombina bombina*) live in small ponds and were given their names because of their yellow and red bellies. If danger is imminent, the so-called 'unken reflex' takes place: the animals throw themselves on their backs and go into a sort of hollow-body position to display the contrasting underside of their bodies. The unken reflex reminds me of the boat position from my yoga practice. While I adopt this position to relax, the information being conveyed in the unken reflex is quite different: with this idiosyncratic posture, the animals are warning their attacker – there is poisonous slime on my belly! Optical warnings in the form of contrasting colours and idiosyncratic postures in animals are often accompanied by acoustic information such as loud hissing, growling, or buzzing. The combination of different communication channels underlines the information content of the threatening postures and makes it clear: not a step closer!

SCREAM IF YOU CAN

Many mammals and birds use acoustic information as an alarm or to make an attacker flee. Before we even set foot in the forest, its inhabitants know that we are there – the call of the jay echoes far through the forest and puts everyone on alert. The information provided by such calls varies according to the type: in members of the squirrel family, such as ground squirrels and marmots, the warning cries pass on information as to how serious the danger is.

The chirping sounds of the Belding's ground squirrel (*Urocitellus beldingi*) are, for example, a signal to fellow members of the species that there's an imminent threat, and that the situation could escalate at any moment. If the ground squirrels emit warbling sounds, then this is purely a signal for increased attention and not a reason to panic yet.

The acoustic signals of meerkats living in Africa (*Suricata suricatta*) not only provide information about the seriousness of the situation, but also contain information about the type of attacker. There's a whole range of predators that may approach from the air and the ground, so it's no wonder that the meerkats have different acoustic signals for these many dangers: the alarm call for an air attack from the martial eagle (*Polemaetus bellicosus*) differs from that for an attack from a jackal (*Canis mesomelas*) and again from the presence of a snake like the cape cobra (*Naja nivea*). The 'snake' call has several different meanings in the 'language' of the meerkats and can also signify traces of faeces, urine, or fur from an unknown attacker nearby — this includes traces of meerkats that aren't part of their own group.

Monkeys like the Ethiopian green monkey (*Cercopithecus aethiops*) also differentiate between dangers in their alarm calls. The call for 'air attack' signals them to raise their heads and look out for danger from above, and to seek protection among bushes.

LIES FOR LIFE

When it comes to survival, it's understandable that a lot of lying takes place. The saying 'all's fair in love and war' seems to fit the predator–prey relationship. Which one of us would not stray from the path of virtue if 'pretending' could save our life? Prime movers in propagating fake news are members of the hoverfly family. The hoverfly, which is harmless, takes on the appearance of defensive wasps, hornets, or bees and bravely buzzes around the enemy's nose, rapidly flapping its wings. The hoverflies rely on the visual information that their role models send out, which signals: 'Get off, I'm poisonous!' Depending on the species, this imitation – mimicry – has varying degrees of success: while particularly large species of hoverflies perfectly imitate their poisonous role models, smaller members of the species don't do it quite so well and portray only a 'cheap' copy of the original. Such differences in the quality of the imitation are best explained by the fact that predators prefer a big meal and therefore mainly focus on the larger hoverflies. That means the danger of being eaten is lower for small hoverflies and therefore so is their need to be particularly convincing in the masquerade.

The possum is also deceptive as it uses the strategy of ... 'playing possum' – pretending to be dead can be life-saving. Many predators react only to fidgeting prey and simply ignore lifeless animals. The possum, or opossum, has impressively mastered this survival trick. It lives in America and is not to be confused with the Australian possum. If the cat-sized common opossum (*Didelphis marsupialis*) is in real danger – as when a predator grabs and starts shaking it – it starts the dying scene: it rolls itself up with open eyes and sticks out its tongue. The opossum can remain motionless in this position for several hours. Once the danger has passed, it rises from the presumed dead and carries on as if nothing has happened.

BEST NOT TO ATTRACT ATTENTION

Active defence mechanisms are one way to defend oneself, but often it's already too late by that point, when the predator has discovered the prey. Not attracting attention is therefore a good plan, and not only for possums. Effective camouflage is everything, and I'm amazed by the accuracy of many animals when it comes to colour mixing. In particular, reptiles, amphibians, and fish are masters of the masquerade and often adapt to their surroundings as if they are part of it. This method of imitating one's habitat is called 'mimesis'. It becomes perfect when colours and shapes can be combined with movements. At first sight, the jerky movements of a chameleon may appear like an unconventional dance performance, but, in its habitat of windblown plants, this visual information blends right in. The Hercules beetle (*Dynastes hercules*) is a further example of how quickly animals can change their colour and merge with their environment. As long as the sun shines, the beetle, which can be up to 17 centimetres long, is green and perfectly suited to its habitat: the forests of North and South America. When it starts to rain, small structures on its body surface take on water and change their shape. This change of shape results in the incoming light refracting differently. Now a different wavelength is reflected: in this way, the colour of the Hercules beetle changes from green to black. So why these colour changes? As long as the sun shines, the beetle's habitat, the forest, is nice and green. However, when clouds appear, the forest darkens and with it the Hercules beetle — it could definitely get a job as a weather presenter!

The last story on the subject of 'eating and being eaten' leads us from America back to Germany, into the Helmholtz Centre for Ocean Research in Kiel.

WHY THE OCTOPUS WOULD RATHER BE A FLATFISH

The young flatfish were piled up in the large water tanks of the research station like whoopee cushions. A good friend and colleague was completing his doctoral thesis here in northern Germany, and I took the opportunity to visit him. 'I just have to feed my fish quickly,' he called out as we made our way to the harbour. My biological curiosity wanted to get up close and personal with the subject of my friend's research, which carried the Latin name *Scophthalmus maximus* – the turbot. Up until this moment, I had only come across this fish in smoked form on my plate. Flatfish have a very individual body shape – the sight of them makes my neck ache. Both eyes are located on the left flank, while the other half of the fish lies flat on the ground. The side that faces upwards looks like it is covered in small stones. The turbot's appearance is optimally adapted to the sandy seabed and thus escapes the eyes of other predatory fish.

What helps the flatfish also seems to work well for the Lilliput longarm octopus (*Macrotritopus defilippi*), which lives on the seabed of the Caribbean. Scientists filmed the octopus, which not only looks like a flatfish but behaves like one too. The animals swim close to the seabed, allowing their eight legs to stream flatly behind them; like a flatfish, they also move forward with jerky movements. In fact, they're almost better at appearing to be a flatfish than the original. The following drawing, however, is of a 'real' flatfish.

Both eyes of the turbot (*Scophthalmus maximus*) are located on the left half of its body. The other half of the fish lies flat on the ground and is perfectly adapted to the stony habitat of the seabed.

Let's get it on?

THE MALE MANY AND THE FEMALE FEW

In the animal kingdom, females prefer to rely on quality rather than quantity: they produce very few, but very large gametes called 'oocytes' and thus invest quite strongly in their potential offspring right from the start. Males, on the other hand, produce a lot of sperm cells, but these are much smaller than the oocytes. Hormones control the maturation of the gametes and ensure that the offspring come into the world at a time of the year when a lot of food is available. This means the females are usually fertile only at certain times, while the males are theoretically always ready to mate. As a rule, the males invest much less time and energy in rearing the young.

In summary, reproduction is a more costly matter for females than for males – and so we are in the middle of the battle of the sexes. Males and females see the issue of reproduction from two different angles and have different motivations for it. Females are very picky when it comes to choosing a father for their future offspring. They have only a limited number of eggs, and of course only the 'best' males will be considered for fertilisation. If the father is healthy and therefore attractive, the probability is high that the offspring will also be highly rated on the 'marriage market' later on. In contrast, males do not have to economise with their sperm, as they have plenty. The more eggs their sperm successfully fertilise, the more offspring they bring into the world. As there are far more sperm cells than oocytes, competition arises among the males – demand simply exceeds the supply. And so, the males court and fight for the females' favour. But they must be sure that the signals they are sending out reach the correct recipient. In other words, in nature, how does a male or a female recognise that communication in matters of reproduction is taking place with the desired recipient of the same species?

ARE YOU THE ONE FOR ME?

Well, how do you know you're dealing with a human being and not a sheep or a goat? At some point, you learned what a person looks like and that you are also one. When you see a fellow member of your species, you recognise his or her human features and the way humans move. It's not very different for other living creatures.

At first sight, the courtship behaviour of birds, fish, or non-human mammals appears to consist of random movements – but this 'sign language' is following a precise pattern. The courtship behaviour takes on an important role in bringing together what belongs together. With their courtship behaviour, the males specifically target females of the same species and often display certain gestures that are supposed to put the female in the mood. Depending on the type of animal, there are different types of behaviour in courtship – ranging from throwing back the head in birds, to positioning the tail in the monkey, right up to zigzag swimming in fish. Seahorses initially 'hold tails' before they mate. With their tail ends tightly entwined, the male and female saunter across the seabed – a sign that they're 'serious'.

The intensity and duration of the courtship and all that goes with it tells us quite a lot about the male's physical strength. Those who invest a lot of time in flirting have to be especially efficient when it comes to foraging or fending off enemies. Multitasking seems to appeal to the female and testifies to the qualities of a real dream male. But it isn't always easy to recognise one's own species, as the example of the Californian deep-sea squid (*Octopoteuthis deletron*) shows us.

DEEP-SEA SQUID TAKE WHAT THEY CAN GET

Male deep-sea squid have a slight problem when it comes to reproduction: their habitat, at a depth of between 400 and 800 metres, is so dark that they can't use any optical information when it comes to choosing a partner. So how do the males recognise the sex of the squid that happens to be swimming past them? Researchers from the American Monterey Bay Aquarium literally brought this out into the light of day. Small deep-sea robots filmed the *Octopoteuthis deletron* on the Californian coastline as they selected their partners. The recordings showed that the males didn't even bother to check the sex of their counterpart: they simply seized every opportunity that appeared in front of their tentacles. Male deep-sea squid leave sperm residue on squid of the opposite sex – but not only of the opposite sex. The researchers also found these typical sperm traces on other male squid – such a high amount that it could no longer be considered an accidental coincidence. This level of indiscriminate mating is an exception in nature, because even the production of cheap sperm cells costs somethings, as does the act of mating itself. In the case of the deep-sea squid, the researchers presume that the accidental mating is the lesser evil in an otherwise extreme habitat. Time is of the essence because finding food and also a suitable mating partner aren't that simple at a depth of 800 metres. If two squid do happen to meet in the dark expanses of the sea, then the male squid can't waste any time trying to find out what sex his counterpart is. This is especially true of the *Octopoteuthis deletron*, because this species has a short life span and reproduces only once.

We don't have to travel to the deep sea to find more examples of such involuntary mating. Male bedbugs (*Cimex lectularius*) are also not choosy when it comes to selecting sexual partners. Once again, it's the traces of mating by males on fellow members of the species that lets on who they spent the night with. Female bedbugs have a small swelling on the underside of their abdomen, which is penetrated by males with a penis-like organ. In this way, the sperm of the male bedbug goes directly into the female's body. As a result of this idiosyncratic manner of reproduction, a wound develops at the penetration point. Yet it's not only the females that develop these wounds. Male bedbugs aren't spared this so-called 'traumatic insemination' by other males. I suppose that's gender equality – of a sort ...

WHAT WOMEN REALLY WANT

Before a female decides on a male, he first has to show that he will be a good father for their joint offspring. But how can the female tell if her chosen one is a good catch? First of all, appearance matters, because physical condition provides important information about the quality of the male, and, in most cases, this isn't easily faked. Body size seems to be a good sign of his health and strength: large males with a lot of testosterone are more aggressive and therefore good protectors of the joint 'home' and the future offspring. So it's no wonder that large males are particularly popular with the 'ladies'. Shiny feathers, a clean coat, or healthy-looking skin are other signs that the male is in the best of health and can take care of himself: if we are ill with the flu, for instance, most people around us will spot this immediately.

Finding a healthy partner increases the chances that the offspring will also inherit a good 'genetic blueprint' and live a long life in order to reproduce. Thus, males don't always send out honest signals when it comes to courting females, using visual aids to adorn themselves with borrowed plumes. Females are therefore well advised to form their own opinions as to the suitability of the future father of their children. Some females go as far as eavesdropping on the males in their everyday life. Female red swamp crayfish (*Procambarus clarkii*), for example, watch as the males fight to access them. When the males present themselves to the females afterwards, the females more often choose the winner of the duels.

SUCCEEDING WITH ALL YOUR WORLDLY BELONGINGS

Many insects and birds bring an edible gift to the courtship as a sign to the females that they're able to procure good food. The male tent spider (*Pisaura mirabilis*) artistically wraps small flies in its gossamer and presents this package to the female on their 'first date'. The bigger the gift, the higher the probability that the male will get lucky. But woe betide the male if he dares to make advances without bringing a gift. In behavioural experiments, biologists at Aarhus University in Denmark observed that the females didn't hesitate to impose the death penalty for bad manners. They ate males without a present much more often than males that satiated the female's hunger with a fly meal before sex. Particularly clever males offer the female a present and then go completely rigid, faking their own death. Once the female starts tucking into the edible gift, the male seizes the opportunity to 'rise from the dead' and get into the right mating position on top of the female. Using this surprise tactic, 89 per cent of the males had sex – but it's hardly a romantic approach.

The example of the bullfrog (*Rana catesbeiana*) shows that females are often not satisfied with just a small gift. They want males with lovely homes! The ideal place to rear bullfrog eggs is warm water without too much vegetation. The frog's eggs develop well here and are protected from predators like the leech. Such ideal habitats are much sought after among the males, and only those that assert themselves get the most desirable property. If a bullfrog manages to call a lukewarm puddle his own, then it won't be long before a female joins him! With his strong croaks, he can direct the female to his manorial pond from a distance of up to two kilometres. The bullfrog owes his name to these acoustic signals, which he uses to attract the female. His croaks are quite low in tone and comparable to the bellowing of an ox. The male bullfrog also uses his eponymous call to keep his rivals at bay: after all, who wants to lose their valuable property to a competitor?

A BOWER FOR THE LADIES

Male bowerbirds are experts when it comes to chic properties. These members of the sparrow family are found in Australia and New Guinea and owe their names to their bowers, which the males build specially to court the females. The females choose which male they want to mate with on the basis of the quality of the bower. No wonder that so many males really go to town when it comes to decorating their love nest. Male bowerbirds zealously select furnishings in suitable colours for the female. This can range from red berries to Coke cans. The males even coat the walls of the bower in 'paints' that they obtain by chewing plants and then apply using their feathers. The females of the different species have different preferences when it comes to colour. One might be delighted by blue, while others prefer bowers in green or red.

The Australian satin bowerbird (*Ptilonorhynchus violaceus*) first builds two parallel walls made of twigs. These walls line a sort of courtyard that can be up to 30 centimetres long, at the northern end of which the bird builds a platform of twigs. The platform now displays the male's entire creativity – he decorates it with a multitude of things, using feathers, flowers, or even snakeskin to make the bower shine in grand style. The aim is to be one of the most impressive house builders in the neighbourhood and to 'pull' as many females as possible and get them to mate. So the males busily work on their bower and defend it against other male bowerbirds. This is necessary as the construction industry of bower love nests isn't always fair. Many a male gets carried away, destroying the bower of a competitor in a cloak-and-dagger action and stealing valuable furnishings for his own interior.

The grey bowerbirds (*Ptilonorhynchus nuchalis*) go so far as to use the power of optical deception to make themselves and their bower appear even bigger from the perspective of the female. They position up to several hundred objects, such as stones and bones, in a special way in the courtyard of their bower. But there won't be long-lasting happiness between the lovers. After mating has taken place in the bower, that's it! The female flies away after fertilisation and builds her own nest elsewhere to raise her chicks.

THE LOUDER THE BIGGER

Australia is also where our next story takes place. Many animals use acoustic signals to convince females of their quality — including the koala (*Phascolarctos cinereus*). Large males send out especially deep calls with which they attract the attention of the females. The female koalas can then choose among the males from a distance, and they usually choose particularly large individuals. The koala male's voice power has long challenged biologists because its calls are comparable in depth and volume to those of a bull elephant. Yet neither the koala's vocal cords nor its larynx is the size of an elephant's — so what's its secret?

The biologist David Reby and his colleague Benjamin Charlton at the University of Sussex, together with researchers from the Leibniz Institute for Zoo and Wildlife Research in Germany, found the solution to the riddle in the koala's nose. They discovered a skin flap that we humans know only too well because it's responsible for our nightly snoring sounds. The male koala, on the other hand, uses this flap of skin — also called the 'soft palate' — to amplify his calls. When calling, the koala simply lowers his larynx and thereby stretches two folds of skin on the soft palate. These folds of skin now cause incoming air to vibrate like two strong vocal cords. The result is a deep sound between 10 and 60 Hertz, which would turn any bull elephant green with envy.

Let's stay with the acoustic signals for a moment but leave Australia behind. In Europe, we come across the partridge (*Perdix perdix*), a small brown bird whose mating season begins in February/March. The partridge is also a good example of how acoustic signals contain information about the quality of their transmitter. Partridge males with higher testosterone levels send out longer mating calls than their male counterparts with less testosterone in their blood. Longer mating calls not only increase the chance that they will be heard by the females, but also signal to the female that the transmitter is persistent and strong. In behavioural experiments, the females more frequently chose the male who sent out particularly long courtship calls.

The sedge warbler (*Acrocephalus schoenobaenus*) has a large vocal repertoire. The birds seek out high reeds, from where their song travels particularly well into the distance. The sedge-warbler song contains long verses that are made up of trills, whistles, and even imitations of the calls of other species. Some males are so adept at singing that they combine many different verses into their courtship calls. These musical qualities are clearly well received by the females: the more verses a male warbles, the faster he finds a female.

The partridge (*Perdix perdix*) is a bird species whose mating season begins in February/March. The higher the testosterone levels of the males, the more persistent their calls to the females (shown here).

DIRTY DANCING À LA GREAT BUSTARD

I was able to observe particularly impressive courtship behaviour during my studies. It was a cold April morning, and I was huddled together with fellow students on a high seat in the expanses of Brandenburg, waiting for the great bustard (*Otis tarda*) to make an appearance. The great bustard is a bird species that's on the verge of extinction and at 16 kilograms is the largest bird in Europe still able to fly. The last specimens in Germany can be admired in the Westhavelland Nature Park in Brandenburg. After two hours in the morning cold, we'd almost given up hope of getting a glimpse of the bird that is jokingly referred to as the 'Brandenburg ostrich'. Suddenly, two dots appeared on the horizon, and, indeed, there it was, the great bustard. Under the expert guidance of the park director, we were hoping to get the chance to view a unique spectacle, which takes place every year in spring over fields in Brandenburg. It's even mentioned in a guidebook with the delightful words, 'Where the bustard farts.' Outside the mating season, the birds live together in groups of the same sex, but, in the spring, the males and females come together to mate. The choice lies with the female, and so they fly for several kilometres among the courting males to make their selection. The long journey is worth it because the male great bustard literally turns himself inside out for the ladies. Inconspicuous and well camouflaged in his brown-grey plumage just a moment ago, the male suddenly jerks his wings around and presents his snow-white elbow feathers on the inside. But there's more. He also folds his tail onto his back, showing off the white underside. The final touches are added to the outfit when the male sets up his long beard feathers. He now looks like a big snowball and attracts the attention not only of the female birds, but also of the many interested birdwatchers.

The show starts as soon as the male bird begins to jerk back and forth. The park director informed us that the bird's heartbeat rises dramatically during this manoeuvre. It goes from around 21 beats per minute to up to 490 beats per minute. The courtship display is primarily a visual spectacle, because the males rarely make any sound. If things do get a bit noisier during the courtship, then it might be because the great bustard male has let off a fart in excitement. As sound recordings have shown, great bustards do actually fart during their courtship manoeuvres – a Brandenburg version of *Gone with the Wind*.

The great bustard (*Otis tarda*) males suddenly turn their wings around during the courtship dance and present the white elbow feathers on the inside of their wings to the females. They also display the white underside of their tail feathers.

THE SCENT OF DESIRE

Gone with the Wind is a good transition to the next communication station and the question: how do wild rabbits arrange to meet? Wild rabbit groups have clearly defined social structures, and there are strict rules governing who is allowed to mate with whom. The highest-ranking male more or less has access to all the females in the group, while lower-ranking animals are disadvantaged. At the beginning of the reproductive period, the composition of hormones and scent in faeces and urine changes – a sign for the rabbits that the search for a partner can now begin. The use of latrines rises noticeably during the reproductive season. When a female wild rabbit presents herself at a latrine to show her willingness to mate, the response from a male in the form of a latrine message is not long in coming. When it comes to finding the right partner, latrines aren't only the method of choice for wild rabbits. Members of the gazelle family such as the Arabian gazelle (*Gazella arabica*) and monkeys such as the moustached tamarin (*Saguinus mystax*) also use latrines as a dating site.

WHY FISH LISTEN TO EACH OTHER

Do you remember the live-bearing fish from my thesis? Now we can find out whether the young male carp change their flirting strategy in the presence of another male. Fish live in shoals, and so the choice of partner is rarely made in private. There's always some other member of the species nearby, observing the partner-choosing process. The Atlantic carp (*Poecilia mexicana*) is one of the few exceptions where the males are the ones choosing their partner. However, in a shoal of carp, there's a huge choice of females and the question of which female any given male should select is not insignificant. So the saying 'What's good for the goose is good for the gander' also applies to the male Atlantic carp. The still inexperienced males observe how the more mature males select a female and copy their behaviour. Even if competition is said to be good for business, the males being observed react to such 'eavesdropping' with a diversion tactic and behave differently.

In an aquarium experiment, I allowed the males to choose between a large and a small female. If there were no spectators, most males spent more time with the large female. This choice makes sense because the large females possess more eggs and are therefore more fertile than the smaller females. But the male carp didn't remain faithful to their 'type' when there was another male in the vicinity, observing them. If another male was around, the fish choosing a partner showed significantly more interest in the smaller female. Clearly the male carp are sending out false signals to the observer, wanting to mislead them – but why the masquerade? There are various theories why such 'bystander effects' exist in the animal kingdom. In order to understand the motivation behind such behaviour, we need to take a closer look not at the conflict between the sexes, but at the battle within one sex.

CRAYFISH PEE ON THEIR RIVALS

The fierce battles within a sex can sometimes end in death. In the fight to claim a female, male red deer (*Cervus elaphus*) literally bash their heads together. So that it doesn't have to come to that, males of many species avoid fighting over mates as much as possible. Instead of actually using their weapons, rivals merely demonstrate them or observe one another in order to get an idea of the rivals' ability to fight. The males of the green swordtail (*Xiphophorus helleri*) do exactly that: they spy on the fights of other males and thereby receive information about their fighting strength. In this way, they know in advance, when they next meet their rival, whether to stand up to him or whether they would lose a fight. The

Galician crayfish (*Astacus leptodactylus*) is a further example of how it makes sense to engage in fights strategically and not blindly. So that a meeting between two males doesn't immediately result in a full-blown fight, the two rivals demonstrate their virility by peeing on one another. The urine allows them to analyse the rival's fitness level and, if necessary, to wave the white handkerchief.

If we think back to the Atlantic carp, one possible explanation for thc male changing his mind with regard to his female of choice is that the males want to avoid aggressive confrontations with other males. By simply opting for a less competitive, smaller female, they avoid unnecessary stress. In theory, this explanation makes sense. In practice, I know that the variety of Atlantic carp that resides in dark caves also hides his choice of partner from other males. These cave-living males are a lot calmer in their interaction with one another; their extreme surroundings are stressful enough – they don't need to bash their fish heads together. So why do the males lie?

Male red deer (*Cervus elaphus*) have antlers, which they use to fiercely fight off rivals during the mating season in September and October. The males attract the females to the rutting ground through scent in their urine.

SPERM RIVALRY — TO THE FASTEST GOES THE PRIZE

The extent to which males compete with each other depends very much on the path of the sperm cells to the egg. Many animals that live in or near water, such as amphibians, fish, or annelids, have indirect fertilisation outside the body. Terrestrial animals – including humans, as well as many insects and arachnids – inseminate the egg directly inside the female. So do sharks and some bony fish. The live-bearing toothcarp, like the Atlantic carp, is an exception when it comes to fish. As the name suggests, the females give birth to their young. This requires the sperm to be implanted directly into the female to fertilise the egg. The male carp has a sort of penis called a 'gonopodium'. If the male is close enough to the female, he swings out the gonopodium and tries to insert it accurately into the female's genital opening. If the gonopodium-swinging is successful, the released sperm can make their way directly to the egg cell for fertilisation. If the female mates with several males, a fight for fertilisation breaks out in her body among the different sperm. The quickest and most resilient sperm prevail and hit the jackpot of life. To stop matters going that far and other males mating with their chosen female, the males try to prevent each other from having sex altogether.

Not wanting to share could be another reason why the male Atlantic carp try to conceal their true love interest from the eyes of their rivals. If the observer copies the choice of female, he will choose the smaller female – removing him from the competition for the larger females.

How far males are prepared to go in order to hold a monopoly over the female of their species is shown in the following examples. The male mourning cuttlefish (*Sepia plangon*) puts on an incredible act when courting a female. If a potential rival approaches, the male swims towards the female and shows her his 'male' side. At the same time, the other side of his body, which is facing his rival, takes on the typical colourings of a female. The rival male is misled and doesn't realise his advances are entirely in vain. In this way, the male mourning cuttlefish skilfully divert attention away from their chosen female.

Other animals even go so far as to 'polish' the genital tract of the female before they insert their own sperm cells. For this purpose, many insects have special structures that help them to remove their predecessor's sperm from the female. Many male mammals 'seal' the female's genital opening straight after sex with a drop of mucus. The male European mole (*Talpa europaea*) is from one of the species that ensures its sperm has a head start by using such a mucus chastity belt. To err on the side of caution, some species 'perfume' the female's genitals with an anti-aphrodisiacal fragrance. Male yellow garden spiders (*Argiope aurantia*) even choose suicide in order to keep the female from further love affairs with other males. They die after copulation in an apparently pre-programmed death while they're still in the mating position on the female: the males themselves become a chastity belt that's extremely hard to crack. Male swallows use slightly less drastic actions to win over their love interest. If the male finds an empty nest because the female has just left, he sends out the acoustic signal 'enemy approaching, danger'. This distress call makes the female return quickly to her nest, although there's actually no reason to.

BRINGING ORDER TO ABUNDANCE

Before we come to the last part of this chapter, I'd like to take you on a small excursion into the history of biology. In 1735, the Swedish naturalist Carl Linnaeus published his book *Systema Naturae*. We humans have always been fascinated by the many living creatures on our earth, and Linnaeus' book was the first major attempt to categorise the many animal and plant species that have resulted from the development of sexual reproduction. Carl Linnaeus compared the appearance and anatomy of countless living beings and tried to integrate them into a system based on their differences.

However, life is extraordinarily diverse, and no sexually created being is identical to another (except in the case of identical twins). In fact, two living beings may differ strongly in their external characteristics yet still be from the same species. We can imagine it like this: you can assemble two cabinets according to the same instructions, but then design them so that one is coloured to match the bedroom and the other fits into the living room. Despite the different colour and size, it's still a piece of furniture of the type 'cabinet'. Now imagine how Linnaeus must have felt. He sees all these pieces of furniture in their houses – aka living creatures in nature – for the first time and tries to bring them into a system. Linnaeus didn't have the means to decipher the building plans of the individual living beings; instead, he examined his fellow creatures closely and classified them into the kingdoms 'animals', 'plants', and 'minerals' according to purely external features.

Today, we have entirely different methods at our disposal and can take a look at genetic blueprints, examining in detail how a living being is related to another. In this way, we're now able to find out whether a particular 'father' is actually the biological creator of his offspring. Birds were once considered to be the poster animal when it came to fidelity, but paternity tests have shown that it's not only avian males that cheat. Female birds also have the occasional dalliance and place the proverbial 'milkman's child' in their nest.

Two, three, many — communication in groups

Living together in society isn't that easy: when many living beings meet or live together in one place, worlds collide. One wants this, another wants that – which can quickly lead to disputes about food or mating partners. So even in the best of families, fights can break out.

SWARMS, STATES, AND FAMILIES — THIS IS HOW ANIMALS LIVE TOGETHER

In the animal world, there are many different ways of living. Some animals are loners and only ever meet fellow members of the species when they're looking for a mating partner or using the same food source. Some live among members of their species at certain times and separate from the group at other times, like when animals mature and go their own way. Others always live in large groups and help each other find food and raise their offspring, as in a bee colony.

Anonymous and open associations may form in gatherings of members of the same species that don't know each other personally. There is always a certain distance maintained between these animals, though they may move closer in situations of danger or cold. An advantage of group life is the mutual exchange of body heat – the more animals huddle together, the cuddlier it becomes, even on the coldest night. Flocks of birds and shoals of fish are good examples of such anonymous, open associations.

Bees, termites, and ants live anonymously but in a closed association. Members are born into their insect state and are part of it throughout their lives – in for a penny, in for a pound!

Mammals such as monkeys, wild rabbits, and gazelles live in closed social systems too, but, in contrast to members of the anonymous insect states, individuals in these mammalian tribes know each other personally through individual communication signals. Living together in this way only works with a clear hierarchy, which is decided in open combat. As soon as it is clear who is in charge, threatening gestures are usually enough to keep things ticking along. The members in such closed and individualised groups are therefore not interchangeable with other members of the species – or can be replaced only with great difficulty. Another example of closed and individualised associations is the family in which at least one parent is involved with their offspring – we humans are a good example of this.

Let's look a little more closely at the individual ways of living together, and especially at how communication works within the group.

'IT WAS THE HERRING!'

The individual members of the species within a swarm don't know each other personally – which makes it all the more astonishing when we see, in videos or with our own eyes, how harmoniously flocks of birds or shoals of fish move as an apparent whole. Visual information such as colour and movement serves to keep the swarm together. As the following story from the secret files of the Swedish Navy shows, fish even use acoustic information to coordinate themselves.

In 1993, underwater sounds presented the Swedish Navy with a great mystery, and this even became a matter of state security. A Swedish submarine repeatedly received suspicious signals on its sonar – a device that uses soundwaves to detect objects underwater. The crew was in agreement: it could only be a Russian submarine! Time and again, the Swedes received the mysterious underwater sounds, but they were unable to find the location of the alleged Russian submarine. In the search for the answer, the Swedish Navy even asked two marine biologists for advice – a good idea, as it turned out. The biologists eventually found out what the signals were. Surprise, surprise: it wasn't the Russians. Far from it! Large swarms of Atlantic herring (*Clupea harengus*) were releasing air bubbles, which confused the Swedish submarine's sonar. However, the biologists were not allowed to publish these amazing findings until years later, because they were sworn to silence over the 'herring case'.

Canadian and Scottish scientists were later able to discover exactly how the herrings produce so many air bubbles and why. In human society, releasing 'wind' in public is likely to lead to ostracism. Not so with herrings. The animals intentionally create gases and allow them to escape to keep the swarm together. Contrary to what was written in the press about this case, the escaping gases from the herrings were not digestive gases, so this means of underwater communication isn't actually 'farting'. In fact, the herrings actively pump air from their swim bladder into the anal tract and in this way produce pulsating sounds. At around 22 Hertz, the herring pulses are emitted in rapid succession and can be heard for a full eight seconds. In contrast to other fish, herrings have body parts in addition to their swim bladder that amplify acoustic information and pass it on to the inner ear. This allows them to hear relatively well. So a function of the shock waves generated by the air pulse could be to provide the swarm's coordinates during the night when visual signals fail. And in future, 'It was the herring!' might be a plausible excuse for underwater flatulence that leads to pronounced bubble formation on the surface.

SAYS ONE BEE TO THE OTHER

One of my parents' neighbours is a beekeeper, and, every summer, his bees buzz around our garden. While my mother is happy that they pollinate her flowers and fruit trees, my father is regularly stung by the bees. As compensation for the pain and suffering, the neighbour regularly gives him a jar of honey. When I consumed this sweet treat as a child, I often wondered how the honey bee (*Apis mellifera*) finds its own food. During my studies, I got a detailed answer and was once again flabbergasted by the ingenuity of nature when it comes to communication.

As in a nation-state, there's a real division of labour among bees, with the scout bees having the task of heading out and discovering the sweet nectar of flowering plants. If a bee has found what it's looking for, it takes a sample of the food source and returns to the hive with this information. Now it is a matter of convincing the bees in the 'collection department' that it's worth flying to the scouted location. How would you explain to your friends that there are rich pickings to be made if you can't use your voice? Body language is the key to success here! If the delicious blossom is nearby, under 100 metres away, the scout bee performs a round dance. She circles to the left and the right. The livelier and more energetic the circles are, the richer the booty is. As proof, the scout bee has the sample in her backpack. If the food source is further away, the scout bee changes her dance. Instead of the round dance, she does a tail-wagging dance (waggle dance) that's reminiscent of a figure of eight. With this far more elaborate performance, she provides information about the direction and distance of the food source.

Bees use not only visual but also acoustic information. The bee's wing movements create mechanical vibrations with a frequency of up to 200 Hertz. The insects really vibrate during their waggle dance — and also during their cleaning dance. In the latter, a bee begins to drum its feet and to shake. It tries to clean its wings. In contact with another bee, the mechanical vibrations are transmitted, and the neighbouring bees also receive the shaking signal. What's the purpose of this shaking process? The vibrations allow the bees to work each other up and motivate themselves for behaviour such as mutual body care. Lucky that we humans have our voices. But maybe a shake-up dance would also be a delightful way of communicating.

THE SCENT THAT SAYS 'KEEP OUT!'

Let's finally solve the mystery of how animals communicate by means of communally used toilets — latrines. Animals living in a group usually share an area where there's food and protection from enemies. This area, their territory, is defended against intruders by all members of the group. Badgers, gazelles, and wild rabbits use latrines to mark out their territory, visually indicating where their own home ends and that of another group begins.

In field studies of different species, it was noticeable that particularly large latrines were found on the borders of territories. These border latrines were patrolled by the highest-ranking males, especially during the mating season, and the males regularly refreshed the latrines with intensive marking. This extra effort on the latrine front pays off because, during the mating season, the males don't only compete for access to the females. The loss of their own territory to an outside competitor is also at stake. We can imagine these latrines as large 'No trespassing' signs along private property. In this case, the message being sent from the mammalian latrine is directed towards fellow members of the species that don't belong to the group and might come up with the idea of trying to nab themselves a territory that's not theirs.

Installing such scented fences does initially come at a cost, but this pays off several times over, as the 'fence latrines' are a means of vivid but non-violent communication. Through their size and smell, they signal the seriousness with which the territory will be defended if necessary. Before an intruder blindly enters into a fight with the owner of a territory, he can, for example, gauge the level of testosterone in the latrines to assess whether he's equal to the highest-ranking male in the group or will come up short.

THE SCENT THAT SPELLS 'SAFETY'

The use of latrines can even save lives, as demonstrated in the Grimms' fairytale of the wolf and the seven little goats. In the story, the wolf tries to convince the goats to open the door to him. The wolf has learned from two failed attempts, and so, in the third attempt, he imitates the high-pitched voice of the nanny goat. He also changes the colour of the fur on his paw from deep black to goat white. A sweet voice and white hooves are the key signals for the goat children to recognise their mother. And so they fall for the wolf's cunning trick and unlock the door, and misfortune takes its course. The wolf gobbles up six of the kids. Yet thanks to one particularly clever little goat, everything ends well: the six young goats are saved and the wolf is drowned. My point is that: the seven little goats would have saved themselves a lot of trauma if only one of them had asked the visitor at the door for a fresh sample of urine or faeces! The wolf in goat's clothing would have had his cover blown immediately, as such individual scents are difficult to fake, even in a fairytale.

In real life, many animals use chemical information for identification. And what could be better as a scent business card than a fresh latrine? Unlike fence latrines, the central latrine of a wild-rabbit territory is used by all members of the group – even the young animals that are only a few months old. Not only does each member leave behind their own personal scent, but the smells mix together into a distinctive 'warren pong', which all members of the group take on automatically when they use the central latrine. This is also the reason why young wild rabbits roll in the latrines in their home territory. If the warren pong hasn't adhered to their fur, then theoretically they don't belong to the group. The pong even seems to trigger a feeling of security and belonging in the adult animals. This central latrine, close to the warren, allows all members of the group to be up to date with all important information on every individual – such as their social status or willingness to mate. The idea of 'shitting at home' thus fulfils the important function of communication within the group – rather like the way we catch up with our regular fellow drinkers in the pub in the evenings or with our colleagues at the coffee machine in the office.

MANAGEMENT SKILLS FOR BADGERS!

Speaking of the office, what does the construction of a latrine communication network have to do with corporate management? I would say quite a lot! Just like us humans, animals also have a certain amount of resources such as time and energy at their disposal. The setting up of a latrine network in particular requires badgers or wild rabbits to confront numerous decisions of economy and time management. I invite you to take part in a thought experiment to show you what I mean.

Imagine that you're a badger and need to set up a communication network of latrines. Get yourself a pen and a piece of paper and take a moment to think. You need to draw the badger's sett as a cross in the middle of the paper. Now draw a circle around the sett: this is the border of your territory. There are other members of your badger group with whom you could construct and maintain around 15 latrines. Play around a little with the allocation of the latrines on your piece of paper and distribute the communication centres as you see fit. The aim is both to protect the borders of your territory and to construct central latrines for communication within the group. Which questions do you need to ask yourself?

Let's assume the following situation: your building plot is located in a popular area, the property market is highly competitive, and there are many interested parties. It's vital to secure the territorial boundaries – you don't want to be continually bothered by unannounced visitors. So should you place all the latrines on the borders of the territory? But what would this mean for internal communication? How many latrines are required for group members to be able to communicate with one another? Keep in mind that the central latrines place a stamp on your territory that reads 'hands off – mine'. This includes the badger's sett as well as the particularly high-yielding feeding places.

Once the latrines have been set up, they need to be checked regularly and freshened up. As mentioned, your property is very large and every journey from the burrow to the edge of the territory requires a lot of time. This raises the cost–benefit question: how many latrines are needed on the borders to protect your own property against a hostile takeover? The more latrines that are set up and need maintaining, the less time you have for activities such as foraging or mating. And who wants to swap an hour of servicing the toilets for a quiet lovers' rendezvous? So you see – the setting up for a latrine network is anything but trivial and requires genuine thought.

No wonder that mammals use a variety of latrine strategies. The spotted hyena (*Crocuta crocuta*) creates relatively small territories and marks the boundaries with numerous latrines. The brown hyena (*Hyaena brunnea*), on the other hand, makes very large territories and deploys the 'hinterland tactic': instead of setting out many latrines on the lengthy border, the brown hyena distributes faeces and urine spots within its territory. Computer models show that this distribution of latrines is indeed the most sensible if you want an intruder to come across a 'no trespassing' sign. For the spotted hyena, its own method of marking appears to be a good balance between costs and benefits of the latrine facility.

The European badger (*Meles meles*) uses accumulations of faeces and urine in its natural habitat to mark out the boundaries of its territory as well as important resources within this area. Depending on the size of the territory and the size of the group of badgers, the number of border latrines varies, as does the frequency with which the latrines are renewed.

In the case of badgers, there's often a distance of several hundred metres between the latrines on the borders of the territory and those near the sett at the centre. Researchers from England wanted to find out which communication strategy badgers from different-sized territories with comparable group sizes considered to be the best option. The larger the territory, the more latrines they built. Some badger groups owned an area of more than 80 hectares, almost as large as 60 football pitches. The communication network consisted of up to 70 latrines, most of them designed as scented fences along the borders. It was striking that the researchers found fewer fresh faeces at these latrines compared to those in smaller territories. In other words, the badgers had more latrines in large territories, but these were used less frequently.

The question remains, how do badgers, wild rabbits, and other animals know what the best solution is for their 'communication department'? Once again, I'm flabbergasted by the amount of know-how to be found in nature.

PART III

What if everything changes?

6

When animals leave the forest

The many examples of the successful exchange of information between living beings are the result of a precise coordination of the transmitter to the receiver. What's more, the development of such information networks is definitively influenced by the habitat: where is the receiver? Which channels are available? Which barriers exist that need to be overcome by the information? It thus takes many generations for a well-functioning communication system to become ingrained – but what happens when things change? An important feature of life is the ability to adapt to a constantly changing habitat and therefore to develop. This also applies, of course, to the sending and receiving of information.

WHAT DO ANIMALS NEED TO SURVIVE IN THE CITY?

Wild boar in Berlin, raccoons in Kassel, and dormice in Osnabrück – in the last few years, there've been many reports about wildlife in the cities of Germany. Apparently, it's not only people who are attracted to metropolitan areas – there's also a rural exodus among animals. This is mainly due to humans using the remaining near-natural habitats. Intensive farming and the spread of cities forces wild animals to leave their previously undisturbed habitats and migrate to other areas. In contrast to the open and empty rural areas, cities with their parks, gardens, and green spaces offer a wide range of suitable housing options for various animal and plant species.

It seems that the more 'courageous' species, those that didn't flee from every passer-by, settled in the cities. Ingenuity and flexibility are two of the attributes required to access the many gifts the city has to offer, such as food and nesting and hiding places. For good reason, it's often the same candidates – foxes (around the world), wild boar (in Europe and Asia), or raccoons (in North America, but spreading) – that reach high densities and lead to conflict with us humans in the cities. There are also many exotic plants in our urban habitat that we humans intentionally plant because they're so colourful, even though they don't belong in our regions. From a biological perspective, these exotic plants speak a different language and may be one of the reasons that the relationships between living creatures in a habitat are changing in the long term.

WHY TORONTO'S NEW RUBBISH BINS WERE A WASTE OF TIME

Cities are particularly interesting habitats, where we can observe communication between living creatures. Here, the habitat conditions change more quickly in comparison to near-natural habitats, and, in the long term, it's only those species that can adapt to change that manage to survive in the city. How well some animals are able to do this is shown in the following story about cunning raccoons in Toronto.

These cute animals, which belong to the family of small bears, regularly plunder the bins and feast on the food leftovers that a large city produces on a daily basis. During the night, the raccoons emerge from their hiding places, simply push over the rubbish bin, and take what they need. The next morning, the remaining waste left lying around testifies to the nightly action of these animals. The authorities in Toronto were so sick of this everyday sight on the streets that, at the end of the 1990s, they decided to turn off the raccoons' food source. No one at the time could have imagined how hard it is to outwit these creatures! The city government invested millions of dollars in new bins with a specially designed raccoon safety catch. These rubbish bins had a screw cap, which was additionally secured by two side closing clamps. My husband grew up in the Canadian metropolis by Lake Ontario and still remembers the initiative known in the media as 'raccoon-proof trash cans in Toronto'. The investment of millions of dollars itself turned out to be rubbish because the bins kept the raccoons from their nocturnal pillaging for only a few weeks.

The animals quickly learned how to open the side closing clamps and the cover with the screw top. The first allegedly burglar-proof bins were replaced by a 2.0 version with additional safety catches – but the raccoons managed to deal with them too. Video cameras on the bins showed evidence of the patience and eagerness to experiment with which the animals approached the task. And so, over the years, these animals became real YouTube stars. You can find countless recordings online if you type in 'raccoon opens trash can' – they're in no way inferior to the cat videos that are so popular. So what makes the raccoons such excellent bin-burglars, and what does this have to do with the exchange of information?

Firstly, these small bears benefit from their body shape. Like a sumo wrestler, the majority of their weight is in the lower body area, which gives them a very low centre of gravity. This helps the raccoons to release forces that allow them to move objects that are many times heavier than themselves. Furthermore, like us humans, raccoons have a rotatable thumb, with which they can grab and manipulate objects.

The city of Toronto unintentionally contributed to the fact that its raccoons have become increasingly clever over the years. With each new version of the rubbish bin, the mammal learned something new. It needed only a few clever raccoons to get the hang of it. The other animals in the group just learned from watching their peers and used their observations for their own careers as bin-crackers. Even young animals are taught the art of 'bin burglary' by their mothers.

It therefore came as no surprise when a study published in 2017 on the number of nerve cells in large carnivores concluded that raccoons must be particularly clever. The raccoons have a high density of nerve cells in their brains, a density usually only found in primates, including us humans. In my in-law's neighbourhood, the residents now have small wooden huts, where the rubbish bins are stored and secured with a padlock. The bins in the huts are tightly packed so that they can't be toppled – but who knows whether one day the raccoons will learn how to crack padlocks!

THE STORY OF THE PEPPERED MOTH

The conditions in our cities present other living beings with entirely new challenges when it comes to communication: constant background noise, polluted air, and contaminated soil obstruct the transmission of acoustic, optical, and chemical information. So any creatures living in the city wanting to successfully send information and communicate in this way need to come up with a plan.

One example is the story of the peppered moth (*Biston betularia*): it begins at the time of the Industrial Revolution in the late 19th century. Before the year 1848, this insect, which belongs to the butterfly family, was very common in England. The night-flying peppered moth, up to 55 millimetres long, was given its name because of its light colouring with some dark speckles, which perfectly mimicked the birch trees they sat motionless on during the day. These dark speckles are due to the pigment called 'melanin' – the more melanin, the darker the colouring. In humans too, the amount of melanin determines the colour of our skin.

In 1848, much darker versions of the moth suddenly began to appear in the city of Manchester. Where did these 'black sheep' suddenly come from? A possible explanation was put forward by the teacher and butterfly researcher James William Tutt. He observed that industrial development in England had significantly affected the environment in and around cities. The sulphur dioxide in the air was killing lichens, some of which are found on the bark of trees. The soot from the factories also settled like a black carpet across the country. To Tutt, it was quite clear what was happening: the previously light-coloured peppered moths were no longer adapted to their habitat because the birch trunks were darker from the soot. A darker variant of the moth would be much less noticeable to birds during the day and would therefore be eaten less frequently. His theory fell on deaf ears. Both other butterfly experts and ornithologists doubted that the peppered moth would actually be eaten by birds during the day and that the moth's different colouring would have an influence on its survival.

It wasn't until the 1950s that the geneticist and butterfly researcher Bernard Kettlewell took a closer look at this question and carried out field tests with the peppered moth. He released specimens in two separate areas. The first area was a mixed forest in Birmingham, which had suffered from industrialisation. The other area was in Dorset, with relatively little pollution and with lichen still visible on its trees. Kettlewell released both light and dark moths in both areas in the early morning and returned in the evening to count those that remained. He wanted to prove that the darker variant had a higher chance of survival than the lighter one in an area that was affected by industrialisation. For the experiment, he exploited one of the moths' characteristics: they don't fly around during the day. So if the moths were no longer in their designated place at the end of the day, it wasn't because they had flown away to sit happily on another tree. In a second part of the experiment, Kettlewell marked light and dark peppered moths and released them in the study area in Birmingham. He then caught the insects again using a moth trap, which contains a scent that attracts moths.

The extensive data of his experiment showed that the dark variant had a better chance of survival in the heavily polluted areas of England. The recapture rate for the dark peppered moths was more than twice as high as for the light ones, and the entire release project showed that fewer darker insects were eaten. However, this was by no means the end of his work on the mysterious dark peppered moth. Kettlewell continued his research until his death in 1979, and scientific interest in the insect remained unbroken. Yet there was still no answer to the question of where this new colour variant had actually come from.

It was the methods used in genetics in the 1960s that brought further insights into how exactly the colouring of the peppered moth works. In 2016, scientists from the University of Liverpool published the long-awaited answer. The unusually dark colouring of the peppered moth from the year 1848 originated in a change

– a mutation – in the DNA containing its plan for the pigment melanin. The scientists were even able to date this change to 1819.

The peppered moth (*Biston betularia*) got its name because in its original colour variation it was perfectly adapted to light birch trees (left). Over the course of industrialisation in England, darker versions of the moth developed (right).

THE EARLY BIRD MAKES ITSELF HEARD

Living creatures that primarily communicate via acoustic information have a particularly difficult time in the city. City noise is most noticeable in low frequencies. These low tones particularly affect birds, which also use low tones to communicate. Not only do they have to 'screech' against the noise of the city, but they also have to factor in the surrounding buildings as 'interference' when sending acoustic signals. Concrete surfaces reflect sound in a completely different way from the trees in the forest. So it's not surprising that there are more bird species that sing at higher pitches in the city than in the country.

The way life in the city influences acoustic communication in animals has been particularly well researched using the example of the blackbird (*Turdus merula*) in Zurich. The blackbirds in the city send out louder signals in addition to making themselves heard above the urban noise level through a higher pitch. Robins (*Erithacus rubecula*) in the city of Sheffield use a different tactic. The urban male robins simply wake up earlier than their counterparts in the countryside and start their singing before sunrise. Sheffield is still largely asleep at this time of the morning, with far less city noise. Another reason for the 'early city birds' may be the constant lighting in cities. In their natural habitat, birds adapt their singing behaviour to the increased light as the sun rises. In cities, however, street lighting means that it never really gets dark – so how are the birds supposed to know when it's time to wake up?

WHY CITY BADGERS KEEP THEMSELVES TO THEMSELVES

Life in the city can also indirectly affect the communication of living beings. In comparison to their 'country cousins', foxes, wild boar, and raccoons in cities spend less time foraging for food. The rich food supply also means the animals' grazing areas become smaller. This means that wild animals in cities no longer have to migrate very far to get to food. City animals also 'save time' by getting used to being permanently disturbed by humans and therefore invest less time in flight than in the country. In the city, for instance, European wild rabbits (*Oryctolagus cuniculus*) are less at risk of being attacked by predators. Of course, their natural enemies, such as birds of prey and foxes, can also be found in the city – but they prefer to take advantage of the more easily accessible food leftovers that humans are kind enough to leave behind all over the city.

Defending large territories, foraging together, and finding themselves subject to predation are important reasons why animals put themselves under the stress of group living in natural conditions. But living together with fellow members of the species has its downsides: diseases can spread faster in the group, and the fight to become top of the hierarchy is a persistent headache. Are wild animals that live in cities better advised to be loners because the disadvantages of group life outweigh the advantages?

To answer this question, it's useful to observe social mammals such as wild rabbits and badgers in country and city habitats. In the countryside, European badgers (*Meles meles*) place great importance on a functioning, well-developed information system using latrines. There, the animals work particularly hard to build their 'scented fences' of latrines on the borders of their territory – the protection of these borders against foreign intruders seems to have the highest priority. In two separate studies, scientists from Bristol and Brighton examined the communication behaviour of urban badgers. But in both cities, the search for badger latrines was inconclusive. The scientists couldn't find a single latrine – either on the borders of the territories or at the badger sett itself.

Is it possible that English badgers don't place any value on marking out their territory in the urban bustle? The scientists took a closer look and discovered that the city badgers demonstrated different social structures compared with their country compatriots. Normally, badgers live in groups with very close social ties and together defend large territories containing food sources. The city badgers, on the other hand, tend to have 'loose' relationships and aren't as interested in the company of other badgers. The oversupply of food seems to make coming together in groups for foraging superfluous. To the badgers, Bristol and Brighton must seem like all-night takeaway food shops.

Rabbits at home in the city

The European wild rabbit is one of those 'courageous' and 'flexible' species that have successfully adapted to city life and become a pest for us humans. While the numbers of rabbits in many rural areas in Germany have gone down continuously for decades, in cities like Berlin, Munich, and Hamburg they have increased. Did you know, for example, that wild rabbits were already going out on the town before the unification of Germany in 1990? Totally unimpressed by the heavily guarded Berlin Wall, the mammals simply burrowed their way under the wall. The art installation 'Rabbit Field' by Karla Sachse on the former death strip on Chausseestraße commemorates these secret Berlin escape artists. In Frankfurt, rabbits haven't respected the green belt for the last decade – much to the annoyance of the city, which for years has been trying to reduce the population density by deploying gamekeepers. Now it is finally time to reveal the secret of the wild rabbits of Frankfurt.

ON THE TRAIL OF RABBITS, WITH SPOTLIGHTS AND CLICKER COUNTERS

When I stumbled across the results of the badger study, it was clear: examining the differences in wild rabbits' latrine distribution in the city and countryside is only half the story! I also needed information about how many animals live together in one place and how the wild rabbits build their warrens.

At dusk, when all rabbits leave their warrens, my team and I got started with handheld spotlights. We searched out the small mammals between blackberry bushes in the surrounding starlit countryside as well as in the inner-city green belt beneath Frankfurt's light-polluted night sky. The scientific literature explains exactly how to work out precise numbers; however, I developed my own method to determine the degree of urbanisation in the 17 field study areas. The intensity of human disturbance and what proportion of the study area was built up were among the four values from which I calculated an index for 'urbanity'. The higher the index, the more urban the area where the rabbits were to be found. This allowed me to see if my urbanity index was related to the communication of the wild rabbits.

I was also able to answer my research question thanks to the help of the local gamekeepers, who, equipped with hunting dogs and ferrets, brought along the right noses for sniffing out wild rabbits. Initially, the hunting dogs tracked down all underground dwellings in the field area, giving me the number of warren openings. The gamekeepers closed off all the warren exits with cages, fitted with clicker counters, and then the ferrets were put to work. A clicking sound indicated that a rabbit had fled to the surface and entered one of the wire cages. So besides the number of warren openings, I also got the numbers of current inhabitants of the rabbit warrens.

BIG CITY, SMALL (RABBIT) HOMES

We found that the density of the wild-rabbit population increased continuously as you approached the city from the countryside. An average of 45 wild rabbits were romping around on one hectare of land in front of the old opera house in Frankfurt. In the most rural of all study areas, there was not a single animal – at least on paper! And the use of ferrets led to amazing findings with regard to the animals' social behaviour. In the city centre, the rabbit warrens usually had fewer than six openings. These small warrens were primarily used by pairs or individual animals. In the countryside, on the other hand, the digging activity of the rabbit groups that were increasingly growing in size was taking on impressive dimensions. Here, we discovered warrens with more than 50 entrances and exits, with up to 15 animals living in each warren.

Another difference between urban and rural animals was shown in their activity: while the wild rabbits in the countryside mostly dared to surface only at twilight, the city rabbits, despite human disturbance, were active even during daylight. Once out of the warren, they also invested half as much time in keeping a lookout for predators compared to the country rabbits. In a nutshell, the Frankfurt city rabbits had a lifestyle that couldn't be more different from their rural counterparts: preferably living alone in a small space, they were constantly on the move. So I had to smile when the press published my results on the behaviour of wild rabbits in the city and compared it to classic human singletons living in cities.

WHY CITY RABBITS BUILD MORE 'FENCES' THAN COUNTRY RABBITS

Let's go back once more: how animals use latrines to communicate is dependent on the population density in the area, the size of the group and the territory, and the probability of being captured by predators. All three circumstances changed for my rabbits, and so it was obvious to me that city rabbits must communicate differently from country rabbits. Thus, my students and I set off in search of latrines, and we were spoilt for choice with a total of 3,272 wild-rabbit latrines in 17 study areas in and around Frankfurt – from blackberry bushes and orchards in the country to carefully mown green areas in front of the Frankfurt opera house. In addition to the distance from the latrines to the respective warrens, we were also interested in the number of fresh faecal pellets as a sign of current use.

Back at my desk, my assumptions were confirmed: there were differences in the latrine communication! The more urban the area became, the more often the wild rabbits created latrines at a certain distance from the warren as scented fences. The fence latrines were much bigger and much closer together than the latrines directly by the warren. A possible explanation for this observation is that, between Frankfurt's high-rise buildings, the rabbits are slowly running out of space and the competition for good territory is increasing. So setting up fences is particularly important to city rabbits if they want to remain landowners. At the same time, the significance of communication through latrines near the warren is diminishing because of the small groups. It's in complete contrast to the rural study areas, where the opposite was true. A large number of latrines were set up directly by the warren and only a few at the edges of the territory. Clearly, the frequent marking of the latrines near the warren is important for communication within the group. With a group of up to 15 members, you have to have these levels of internal information exchange. The country rabbits seem to have fewer problems with neighbours, as the next warren is often several kilometres away. In such a relaxed residential area, there's no need for intensive signage of territorial borders.

The European wild rabbit (*Oryctolagus cuniculus*) is a typical example of a wild animal species that has reached high densities in cities. The animals communicate in their group via latrines. Latrines on the periphery of the warren, on the other hand, signal the boundaries of the territory to rabbits outside the group.

HOW ARE THE WILD RABBITS IN FRANKFURT DOING TODAY?

I ended my data collection in Frankfurt in December 2014 and moved back to my old home in Brandenburg. At that time, Frankfurt had a large population of wild rabbits, and I was sure the city gamekeepers would still be dealing with the so-called 'rabbit plague' for years to come. But neither the gamekeepers nor I could have envisaged how quickly things would change: I came to Frankfurt several times over the next few years, and each time I noticed that there were fewer wild rabbits to be seen. In areas where you were previously guaranteed to see large numbers, there were now only single rabbits or none at all to be seen.

In September 2018, I returned to my old field-study area with a camera team to film a documentary about rabbits and their habitats worldwide. In cooperation with the local gamekeepers and accompanied by the camera team, I set off on the trail of the rabbits – and was shocked by what we discovered: the formerly high population density of wild rabbits seemed to have shrunk dramatically! Where I'd counted 20 wild rabbits on the green areas in front of the Frankfurt opera house in 2014, I now counted a total of four. What had happened to Frankfurt's rabbits? Had they all migrated back to the country?

From conversation with the city gamekeepers and interested citizens, two possible explanations emerged: firstly, a new variant of the deadly rabbit haemorrhagic disease (RHD) had decimated rabbit numbers; and secondly, the once favourable habitat for wild rabbits in Frankfurt had undergone a negative change. Wild rabbits prefer hedges and bushes in green areas to create protected warrens. During my doctoral studies, it had at times been impossible to get a close look at the warrens because of the dense vegetation. In 2018, these formerly hidden spaces were laid wide open.

The search for the missing wild rabbits remained unsuccessful for me and the film team in September 2018, and, even today, the city gamekeepers report that the situation is unchanged. The file on 'Frankfurt's wild rabbits' is not closed yet, though – I will continue to gather data on the origin and health status of the animals in the city area. Evaluating and publishing this data will hopefully shed some more light on the situation. In the meantime, I would love to receive any relevant information about the whereabouts of my former study subjects in and around Frankfurt!

And the moral of the story?

At this point, we've almost reached the end of the book. I don't know about you, but, after all these examples about the exchange of information in nature, I'm fascinated anew by the life around me. Ever more precise scientific methods allow us humans insights into a world of information exchange between living beings that we were previously unaware of. So today we can trace the reaction of organisms to incoming information right down to the level of their cells. Let's think back to the natural scientists in the 18th century, a time when scholars classified fungi as inanimate minerals. Now we know what communication achievements these living beings are capable of! When studying the latest research results, I marvel at the sometimes incredibly precise and 'imaginative' communication of protozoa, fungi, plants, and animals. But before I tell you what I've learned about my own communication behaviour from all my years of studying behavioural biology, let's summarise the most important findings.

A WORLD OF DATA

The world out there is full of data – not only for us humans, but also for all living beings. This data becomes 'information' when living beings perceive it through their receptors. Depending on which receptors a living being has, it can perceive different information from its surroundings. This is how the development of unicellular and multicellular organisms goes hand in hand with their habitat and lifestyle. If a living being has eyes, it perceives visual information such as colours, shapes, and movements – and can use these to communicate. Flowering plants, for example, seem to know that their pollinators are able to perceive electromagnetic radiation in the UV range particularly well, and so target the pollinators with striking patterns in these 'invisible' colours that are especially attractive to them.

If a living being wants to actively transmit data to a recipient, it can bundle the data in a transportable package – the signal. This signal transports the data through the living being's habitat to the recipient. When the recipient unpacks the signal, it perceives the data with its receptors, and the data becomes information. A prerequisite is that the sender and the recipient have a common 'data pool' – in other words, that they speak the same language.

PERCEIVING INFORMATION FROM THE SURROUNDINGS – NOW IT'S YOUR TURN!

We humans also perceive our environment by using our receptors – the sensory organs. We see, hear, smell, feel, and taste. In everyday life, this happens quite incidentally, and we're often not aware of the information we digest each day. Do we actually need it all? Perhaps we even miss information in certain areas because we don't have the 'sense' for it? Now I'd like you to explore these questions and to consciously focus on the information you receive.

You'll need a timer, a pen, and 30 minutes of your time. First focus on your surrounding for five minutes, and consciously perceive what you see with your eyes. When the five minutes are over, take five minutes to write down all the visual information you saw, such as shapes, colours, and movement.

What did you see?

Perception	Reaction

For the next five minutes, focus your attention on your acoustic receptors, your ears. What can you hear, and where is the acoustic information coming from? Then write down everything you heard.

What did you hear?

Perception	Reaction

For the last five minutes of observation, focus on your nose. Pay careful attention to what you can perceive from your surroundings, and make a note of everything you can smell.

What did you smell?

Perception	Reaction

Take a look at your notes again, and then write down your spontaneous reaction to each piece of information. Don't think about it for too long – your first impulse is the best.

Here some examples from my information list:

Red dress – provocative
Green tree – relax
Smell of burnt food – annoyance that this happened again
Hissing cat – aggression

The point of this exercise is to show you that we are constantly surrounded by data and perceive this as information. From the point of view of communication, a key factor is which information reaches the recipient and how he or she reacts to it. Certain movements, sounds, or smells might make me angry, but might leave you completely cold. This is also the reason why communication is so susceptible to disruption. Even between people who speak the same language, one and the same word can mean something different and therefore trigger a different reaction depending on the recipient's individual interpretation. Human language is rarely as precise as the key–lock principle of a scent on a chemical receptor. Fortunately, we humans also use such non-verbal means of communication and are able to iron out a verbal slip-up by using a suitable pheromone with regard to the object of our desire. A friendly smile often says more than a thousand words and is understood in all languages.

THE JOYS OF TRAVEL

I often have interesting experiences when I travel – in the confined space of a train or a plane, I quickly strike up conversation with my fellow international travellers. In human communication, social aspects such as culture, tradition, and convention play an important role. Colours and gestures can take on very different meanings. For example, my biggest communication faux pas happened to me at a conference in Sapporo, in Japan: a young Japanese man next to me kept sniffing very loudly throughout the lecture. With a friendly smile, I reached out and offered him a tissue, which to my surprise he turned down with an angry look on his face. I looked around and realised that some of the other conference delegates who had seen what had happened were giving me dirty looks. Although I had studied my travel guide in detail before arriving in Japan, I had forgotten that passing someone a handkerchief was tantamount to something like a declaration of war. Socially, it's more acceptable in Japan to 'sniff' than to 'blow'.

But I needn't travel so far from home to witness such misunderstandings. Travelling by train from Berlin to Frankfurt, I had set up my laptop on a high table in the buffet car, when I witnessed a scene of typical human communication. An older gentleman was waiting at the serving counter and wanted to order a coffee from the lady there.

He: 'A coffee please.'
She: 'To drink here or to go?'
He: 'Don't you have any cups?'
She: 'Yes, so you want the coffee to drink here?'
(A train rushed by and swallowed the lady's words.)
He: 'I didn't hear what you said. I would like a coffee.'
She: 'Do you want the coffee to go?'
(The gentleman's face started to turn red, he began to sweat, and his voice got louder.)
He: 'No, don't you have any cups?'
(The lady frowned. I could see she was irritated.)
She: 'Yes, so you want to drink the coffee here?'
He: 'Of course, where else?'
(She silently passed him the cup of coffee. He paid and even gave her a tip.)

The entire communication between the elderly gentleman and the buffet lady took just minutes. Let's remember that it was only about ordering a cup of coffee. I would venture to suggest that this exchange of information between sender and receiver was rather suboptimal. The gentleman clearly did not know that a 'coffee to go' meant a 'coffee in a disposable cup'. The man was obviously not using the same vocabulary as the woman. The inevitable deterioration in hearing ability in old age combined with the sounds of trains passing by in the background additionally complicated the communication between the two. Sender and receiver in this example were clearly not on the same wavelength and were therefore not in resonance. For the other train passengers, including me, as observers of the situation, it was easy to read between the lines. While both tried to remain friendly and not adopt the wrong tone, their faces and postures were saying something completely different.

THE BEDROCK OF COMMUNICATION — WHAT EXACTLY ARE WE TALKING ABOUT?

'Some conversations are about as useful as two days of roundabouts.'

Anonymous

Who could teach us more about useful communication than the creatures that surround us, whose daily survival is dependent on successful coordination and organisation with countless other creatures in their habitats? Let's think back to the bees. They show full physical commitment when it comes to passing on useful information to their many fellow inhabitants of the hive. We recall how communication is supposed to reduce 'lack of knowledge' by sending and receiving information. In other words – following a conversation with a fellow human being, we should be in some way wiser than before. We can then use this new information as useful knowledge in upcoming decisions in everyday life.

If I encounter communication barriers when dealing with other human beings, I ask myself an important question: what is this actually about? Just like in nature, depending on who we are dealing with, there are different motivations for sending and receiving information. In the win-win situation I already mentioned, both sender and receiver benefit positively from the communication. If you are dealing with someone with whom you want to communicate more than once, it's a good idea to send honest information right from the start. We can expect this kind of communication with relatives or colleagues. The tendency to exaggerate slightly and to be untruthful, especially between the sexes, is another story ...

I get the impression that in some business meetings, the

communication is quite similar to that between predators and prey – do or die. If you want to negotiate your salary with your boss, you probably have different interests in the outcome of the conversation. You, as the 'lower-ranking person', want to assert yourself and not let yourself be deterred. In other situations, we want exactly the opposite – to not attract any attention to ourselves or to even play dead like the possum or merge into the background like the turbot. When my doctoral supervisor asked me about the status of my thesis, I sometimes wished I had those skills.

HOW MUCH INFORMATION DO WE REALLY NEED?

If we know the reason for communication, it's easier to send concise information. Clarity doesn't only help the sender to meet their communication goal more quickly – it also saves the recipient time and stress. Animals and plants in nature don't have time to beat about the bush. Their signals have become so optimised that they send all the necessary information in a short space of time.

But who decides whether information is important or not? In fact, it all depends on the recipient's interpretation. You too can ask yourself what information you need to successfully master the tasks that come up in everyday life. Or the other way around – which information can you do without? The quest for food, reproduction, and the easy life is also on our daily communication agenda. If hunger drives us to a restaurant, the reason for sending the information is clear – food acquisition. However, the selection is often too wide, and so often my reply when asked for my order is: 'I don't know yet. I need a moment.' In most cases, I need 15 minutes to choose what I want to eat. It's obvious: if we ourselves don't know what we want to eat, then we can't communicate this information clearly either.

WHY CONTACT WITH NATURE HELPS US TO COMMUNICATE

When we're relaxed and our minds are clear, we are more capable of seeing what things are important to us and what we want to talk to other people about. When we exchange information with others, it makes a difference whether we're stressed or whether we're physically and mentally balanced. Humans are continuously moving towards man-made environments such as cities, and this naturally places stress on us and on the non-human living beings around us. Studies confirm that people who live in urban areas are particularly prone to stress and that this stress diminishes as soon as we find ourselves in semi-natural habitats such as forests, mountains, or lakes.

A few hours in a forest have an immediate positive effect on our immune and hormone systems. The Japanese even have a special term for it: *shinrin-yoku*, which means 'taking a bath in the atmosphere of the forest', or, put more simply, 'forest bathing'. This type of preventative health measure is generally accepted among the Japanese and has led to a phenomenon called 'forest tourism'.

We find calmness in nature, our thoughts slow down, and we relax. The way I see it, a healthy diet, exercise in the fresh air, and sufficient relaxation aren't only the key to our quality of life – these things also help with communication.

OFF TO THE FOREST WITH YOU

Which story in this book impressed you the most? The sending of light signals in dark caves and at the bottom of the sea? The communication between mycorrhizal fungi and plant roots? Or was it the latrine-building wild rabbits? For me, each of these examples is a testimony to nature's more sophisticated communication strategies, at which we humans can only marvel.

Communication was not our idea but has been the connection between all living creatures since the beginning of time. Look at the way a flower 'knows' that it is more likely to be pollinated if it sends out a certain visual signal.

We often seem to forget that humans are also living creatures and are therefore part of the great whole on this earth. So let's go and bathe in the forest more often and spend more time in nature – and while we are at it, perhaps we should take our family, friends, and bosses with us! That way, we could receive unexpected information that might help us to have new ideas. If that happens for you, share the ideas with the living beings around you. Who knows what amazing insights into the 'language of nature' we might get in the future? This much is clear: every single living thing sends and receives information!

Acknowledgements

From the initial idea to the finished book, I was accompanied and supported by many whose advice and encouragement was invaluable. My deepest thanks therefore go to the following people:

Inga Poste, who saw me at my very first science slam in Berlin and had the idea for the book. Anna Frahm and Catharina Stohldreier from the publisher Piper Verlag, who answered all my first-time-author questions and accompanied me through the adventure called 'a book'.

Stefan Christ, Swetlana Gutwin, Dr Bernd Hermann, Dr Hannes Lerp, Dr Wiebke Ullmann, Albrecht Vorster, and Wolfgang Ziege, who, in some cases, even went through the manuscript multiple times.

Kerstin Bosse, Andreas Fiedler, Jana Freymark, Professor Katja Puteanus Birkenbach, Dr Franziska Schwarz, and Heiderose Ziege for the enlightening conversations over coffee in Brandenburg.

Moving on to the Canadian city of Toronto and Marcus MacDonald and Ted and Mary McIntyre. Marcus was decisive in helping me put together the many puzzle pieces, and the McIntyre family gave me a great deal of support.

My former work group at the Goethe University in Frankfurt as well as all enthusiastic scientists around the world who bring new facts about the processes in nature to light every day.

Last but not least, to my most important companion, Eris Fellmeth. Thank you for your wit, your sharp mind, and your unshakable faith in me.

Bibliography

General

Ahne W, Liebich HW, Stohrer M, Wolf E (2000) Zoologie: Lehrbuch für Studierende der Veterinärmedizin und Agrarwissenschaften, mit 25 Tabellen; Glossar mit 551 Stichwörtern. Schattauer Verlagsgesellschaft mbH, Stuttgart, New York.

Bear MF, Connors BW, Paradiso MA (2018) Neurowissenschaften: Ein grundlegendes Lehrbuch für Biologie, Medizin und Psychologie. 4. Auflage. Springer-Verlag, Berlin, Heidelberg.

Bradbury JW, Vehrencamp SL (1998) Principles of Animal Communication, 2nd Edition. Sinauer Associates, Sunderland, MA.

Campbell NA, Reece JB (2011) Biologie: gymnasiale Oberstufe, Band 4900 von Pearson Schule Pearson Studium – Biologie Schule. Pearson Deutschland GmbH.

Duden (2016) Deutsches Universalwörterbuch: Das umfassende Bedeutungswörterbuch der deutschen Gegenwartssprache. Bibliographisches Institut.

Eckert R, Randall DJ, Burggren W, French K (2002) Tierphysiologie, 4. Auflage. Georg Thieme Verlag, Stuttgart, New York.

Frings S, Müller F (2019) Biologie der Sinne: Vom Molekül zur Wahrnehmung. 2. Auflage Springer-Verlag, Berlin, Heidelberg.

Gruner HE, Kaestner A (1993) Lehrbuch der speziellen Zoologie. Band I: Wirbellose Tiere. Teil 1: Einführung, Protozoa, Placozoa, Porifera. Fischer Verlag, Stuttgart.

Heldmaier G, Neuweiler G, Rössler W (2013) Vergleichende Tierphysiologie: Neuround Sinnesphysiologie. Springer-Verlag, Berlin, Heidelberg.

Kappeler P (2006) Verhaltensbiologie. Springer-Verlag, Berlin, Heidelberg.

Leonard AS, Jacob SF (2017) Plant-animal communication: past, present and future. Evol Ecol 31:143–151.

Maynard Smith J, Harper D (2003) Animal Signals. Oxford University Press, Oxford.

Müller WA, Frings S, Möhrlen F (2019) Tier- und Humanphysiologie: Eine Einführung, 6. Auflage. Springer-Verlag, Berlin Heidelberg.

Poeggel G (2013) Kurzlehrbuch Biologie, 3. Auflage. Georg Thieme Verlag, Stuttgart.

Schaefer HM, Ruxton GD (2011) Plant-Animal Communication, 1st Edition. OUP Oxford.

Seyfarth RM, Cheney DL (2003) Signalers and Receivers in Animal Communication. Annu Rev Psychol 54:145–173.

Sitte P, Strasburger E, Weiler EW, et al (2002) Strasburger — Lehrbuch der Botanik für Hochschulen, 35. Auflage. Spektrum Akademischer Verlag, Heidelberg, Berlin.

Wehner R, Gehring WJ (2007) Zoologie: 17 Tabellen; Glossar mit 830 Stichworten. 24. Auflage. Georg Thieme Verlag. Stuttgart.

Wilczynski W, Ryan MJ (1999) Geographic variation in animal communication systems. In: Foster S, Endler JA (eds) Geographic Variation in Behavior, Perspectives on Evolutionary Mechanisms. Oxford University Press, New York, Oxford.

Witzany G (2013) Biocommunication of animals. In: Biocommunication of Animals. pp. 1–420.

Witzany G (2017) Key levels of biocommunication. In: Biocommunication: Sign-Mediated Interactions between Cells and Organisms. World Scientific, pp. 37–61.

Ziege M, Babitsch D, Brix M, et al (2013) Anpassungsfähigkeit des Europäischen Wildkaninchens (*Oryctolagus cuniculus*) entlang eines rural-urbanen Gradienten. Beiträge zur Jagdund Wildforsch 38:189–199.

Ziege M, Babitsch D, Brix M, et al (2016) Extended diurnal activity patterns of European rabbits along a rural-to-urban gradient. Mamm Biol 81:534–541.

Ziege M, Bierbach D, Bischoff S, et al (2016) Importance of latrine communication in European rabbits shifts along a rural-to-urban gradient. BMC Ecol 16. doi: 10.1186/s12898-016-0083-y.

Ziege M, Brix M, Schulze M, et al (2015) From multifamily residences to studio apartments: Shifts in burrow structures of European rabbits along a rural-to-urban gradient. J Zool 295:286–293.

Ziege M, Mahlow K, Hennige-Schulz C, et al (2009) Audience effects in the Atlantic molly (*Poecilia mexicana*) – prudent male mate choice in response to perceived sperm competition risk? Front Zool 6:1–8.

Zrzavý J, Storch D, Mihulka S (2009) Evolution: Ein Lese-Lehrbuch. 2. Auflage. Springer-Verlag, Berlin, Heidelberg.

Introduction

Billiard S, López-Villavicencio M, Devier B, et al (2011) Having sex, yes, but with whom? Inferences from fungi on the evolution of anisogamy and mating types. Biol Rev 86:421–442.

Eisler R (1912) Philosophen-Lexikon. In: Bertram M (ed) Geschichte der Philosophie. Directmedia Publ., Berlin, p. 22031.

Griffin AS (2004) Social learning about predators: a review and prospectus. Learn Behav 32:131–40.

Huber H, Hohn MJ, Rachel R, et al (2002) A new phylum of Archaea represented by a nanosized hyper-thermophilic symbiont. Nature 417:63–67.

Jahn U, Gallenberger M, Paper W, et al (2008) *Nanoarchaeum equitans* and *Ignicoccus hospitalis*: New insights into a unique, intimate association of two archaea. J Bacteriol 190:1743–1750.

Matsuhashi M, Pankrushina AN, Takeuchi S, et al (1998) Production of sound waves by bacterial cells and the response of bacterial cells to sound. J Gen Appl Microbiol 44:49–55.

Ritchie D (1986) Shannon and Weaver: Unravelling the paradox of information. Communic Res 13:278–298.

Seyfarth RM, Cheney DL (2003) Signalers and Receivers in Animal Communication. Annu Rev Psychol 54:145–173.

Shannon CE (1948) A mathematical theory of communication. Bell Syst Tech J 27:379–423.

Shannon CE, Weaver W (1998) The mathematical theory of communication. University of Illinois Press.

Tembrock G (2003) Biokommunikation: Nachrichtenübertragung zwischen Lebewesen. In: Kallinich J, Spengler G (eds) Tierische Kommunikation, Braus. Heidelberg, pp. 9–27.

Wilczynski W, Ryan MJ (1999) Geographic variation in animal communication systems. In: Foster S, Endler JA (eds) Geographic Variation in Behavior, Perspectives on Evolutionary Mechanisms. Oxford University Press, New York, Oxford, p. 234.

Wiley RH (1983) The evolution of communication: information and manipulation. In: Halliday TR, Slater PB (eds) Animal Behaviour, 2nd edn. Oxford (UK): Blackwell Scientific, pp. 156–189.

Witzany G (2013) Biocommunication of animals. Springer-Verlag, Heidelberg, New York London.

Witzany G (2017) Key levels of biocommunication. In: Biocommunication: Sign-Mediated Interactions between Cells and Organisms. World Scientific, pp. 37–61.

Chapter 1. Life is live

Baluška F, Mancuso S (2018) Plant Cognition and Behavior: From Environmental Awareness to Synaptic Circuits Navigating Root Apices. In: Baluška F, Gagliano M, Witzany G (eds) Memory and Learning in Plants. Springer International Publishing, Cham, pp. 51–77.

Barja I, List R (2006) Faecal marking behaviour in ringtails (*Bassariscus astutus*) during the non-breeding period: spatial characteristics of latrines and single faeces. Chemoecology 16:219–222.

Blackledge TA (1998) Signal conflict in spider webs driven by predators and prey. Proc R Soc London Ser B Biol Sci 265:1991–1996.

Böhle M, Oertel H, Ehrhard P, et al (2013) Prandtl – Führer durch die Strömungslehre: Grundlagen und Phänomene. In: 12. Auflage. Vieweg+Teubner Verlag, Wiesbaden, p. 656.

Bull CM, Griffin CL, Johnston GR (1999) Olfactory discrimination in scatpiling lizards. Behav Ecol 10:136–140.

Dettner K, Peters W (2011) Lehrbuch der Entomologie, 2. Auflage. SpringerVerlag, Berlin, Heidelberg.

Gagliano M (2012) Green symphonies: A call for studies on acoustic communication in plants. Behav Ecol 24:789–796.

Gagliano M, Grimonprez M (2015) Breaking the Silence – Language and the Making of Meaning in Plants. Ecopsychology 7:145–152.

Gagliano M, Mancuso S, Robert D (2012) Towards understanding plant bioacoustics. Trends Plant Sci 17:323–325.

Hesterman ER, Mykytowycz R (1968) Some observations on the odours of anal gland secretions from the rabbit, *Oryctolagus cuniculus* (L.). CSIRO Wildl Res 13:71–81.

Hughes M (1996) Size assessment via a visual signal in snapping shrimp. Behav Ecol Sociobiol 38:51–57.

Kurzweil P, Frenzel B, Gebhard F (2009) Physik Formelsammlung: mit Erläuterungen und Beispielen aus der Praxis für Ingenieure und Naturwissenschaftler. Vieweg+Teubner Verlag, Wiesbaden.

Li Q, Wang J, Sun H-Y, Shang X (2014) Flower color patterning in pansy (*Viola × wittrockiana* Gams.) is caused by the differential expression of three genes from the anthocyanin pathway in acyanic and cyanic flower areas. Plant Physiol Biochem 84:134–141.

Luginbühl P, Ottiger M, Mronga S, Wüthrich K (1994) Structure comparison of the pheromones Er-1, Er-10, and Er-2 from *Euplotes raikovi*. Protein Sci 3:1537–1546.

MacGinitie GE, MacGinitie N (1949) Natural History of Marine Animals. McGraw-Hill Book Company, New York.

Matsuhashi M, Pankrushina AN, Takeuchi S, et al (1998) Production of sound waves by bacterial cells and the response of bacterial cells to sound. J Gen Appl Microbiol 44:49–55.

Mykytowycz R (1968) Territorial marking by rabbits. Sci Am 218:116–126.

Mykytowycz R (1974) Odor in the spacing behaviour of mammals. In: Birch MC (ed) Pheromones. Amsterdam: North-Holland, pp. 327–343.

Mykytowycz R, Gambale S (1969) The Distribution of Dung-Hills and the Behaviour of free living Wild Rabbits, *Oryctolagus cuniculus* (L.), on them. Forma Funct 1:333–349.

Mykytowycz R, Hestermann ER (1975) An Experimental Study of Aggression in Captive European Rabbits, *Oryctolagus cuniculus*. Behaviour 52:104–123.

Ritzmann RE (1974) Mechanisms for the snapping behavior of two alpheid shrimp; *Alpheus californiensis* and *Alpheus heterochelis*. J Comp Physiol 95:217–236.

Schein H (1975) Aspects of the aggressive and sexual behaviour of *Alpheus heterochaelis*. Mar Freshw Behav Physiol 3:83–96.

Schön Ybarra MA (1986) Loud Calls of adult male red howling monkeys (*Alouatta seniculus*). Folia Primatol 47:204–216.

Takahashi H, Suge H, Kato T (1991) Growth promotion by vibration at 50 Hz in rice and cucumber seedlings. Plant Cell Physiol 32:729–732.

Tuxen SL (1967) Insektenstimmen. Springer-Verlag, Berlin, Heidelberg.

Versluis M, Schmitz B, Von der Heydt A, Lohse D (2000) How snapping shrimp snap: Through cavitating bubbles. Science 289:2114–2117.

von Byern J, Dorrer V, Merritt DJ, Chandler P, Stringer I, Marchetti-Deschmann M, McNaughton A, Cyran N, Thiel K, Noeske M, Grunwald I (2016) Characterization of the fishing lines in Titiwai (= *Arachnocampa luminosa* Skuse, 1890) from New Zealand and Australia. PLoS One 11, e0162687.

Wronski T, Plath M (2010) Characterization of the spatial distribution of latrines in reintroduced mountain gazelles (*Gazella gazella*): do latrines demarcate female group home ranges? J Zool 280:92–101.

Zollner PA, Smith WP, Brennan LA (1996) Characteristics and adaptive significance of latrines of swamp rabbits (*Sylvilagus aquaticus*). J Mammal 77:1049–1058.

Chapter 2. Life is on stand-by

Blaxter JHS, Denton EJ, Gray JAB (1981) Acousticolateralis system in clupeid fishes. In: Tavolga WN, Popper AN, Fay RR (eds) Hearing and Sound Communication in Fishes. Springer-Verlag, New York, pp. 39–56.

Boistel R, Aubin T, Cloetens P, et al (2011) Whispering to the deaf: Communication by a frog without external vocal sac or tympanum in noisy environments. PLoS One 6:e22080. doi: 10.1371/journal.pone.0022080.

Cator LJ, Arthur BJ, Harrington LC, Hoy RR (2009) Harmonic convergence in the love songs of the dengue vector mosquito. Science 323:1077–1079.

Eckert R, Randall DJ, Burggren W, French K (2002) Tierphysiologie, 4. Auflage. Georg Thieme Verlag, Stuttgart, New York.

Ehret G, Tautz J, Schmitz B, Narins PM (1990) Hearing through the lungs: lung-eardrum transmission of sound in the frog *Eleutherodactylus coqui*. Naturwissenschaften 77:192–194.

Fiedler K, Lieder J (1994) Mikroskopische Anatomie der Wirbellosen. Gustav Fischer Verlag, Stuttgart.

Glaeser G, Paulus HF (2014) Linsenaugen oder Facettenaugen. In: Glaeser G, Paulus HF (eds) Die Evolution des Auges – Ein Fotoshooting. Springer Spektrum, Berlin, Heidelberg, pp. 16–59.

Hase A (1923) Ein Zwergwels, der kommt, wenn man ihm pfeift. Naturwissenschaften 11:967.

Hetherington TE (1992) The effects of body size on functional properties of middle ear systems of anuran amphibians. Brain, Behav Evol 39:133–142.

Kurzweil P, Frenzel B, Gebhard F (2009) Physik Formelsammlung: mit Erläuterungen und Beispielen aus der Praxis für Ingenieure und Naturwissenschaftler. Vieweg+Teubner Verlag, Wiesbaden.

Ladich F (2013) Akustische Kommunikation bei Fischen: Lautbildung, Hören und der Einfluss von Lärm. In: Sitzungsberichte der Gesellschaft Naturforschender Freunde zu Berlin. pp. 83–94.

Lenz P, Hartline DK, Purcell J, Macmillian D (1995) Zooplankton: Sensory Ecology and Physiology. CRC Press.

Lindquist ED, Hetherington TE, Volman SF (1998) Biomechanical and neurophysiological studies on audition in eared and earless harlequin frogs (Atelopus). J Comp Physiol A Sensory, Neural, Behav Physiol 183:265–271.

Lindquist ED, Hetherington TE (1996) Field Studies on Visual and Acoustic Signaling in the 'Earless' Panamanian Golden Frog, *Atelopus zeteki*. J Herpetol 30:347–354.

Mischiati M, Lin HT, Herold P, et al (2015) Internal models direct dragonfly interception steering. Nature 517:333–338.

Miyoshi N, Kawano T, Tanaka M, et al (2003) Use of Paramecium Species in Bioassays for Environmental Risk Management: Determination of IC50 Values for Water Pollutants. J Heal Sci 49:429–435.

Montealegre ZF, Robert D (2015) Biomechanics of hearing in katydids. J Comp Physiol A 201:5–18.

Montealegre-Z. F, Jonsson T, Robson-Brown KA, et al (2012) Convergent evolution between insect and mammalian audition. Science 338:968–971.

Neuweiler G, Heldmaier G (2003) Das Seitenliniensystem. In: Vergleichende Tierphysiologie: Neuround Sinnesphysiologie. Springer-Verlag, Berlin, Heidelberg, pp. 199–209.

Plath M, Parzefall J, Körner KE, Schlupp I (2004) Sexual selection in darkness? Female mating preferences in surface- and cave-dwelling Atlantic mollies, *Poecilia mexicana* (Poeciliidae, Teleostei). Behav Ecol Sociobiol 55:596–601.

Schmidt RF, Lang F, Heckmann M (2007) Physiologie des Menschen: Mit Pathophysiologie. 30. Auflage. Springer-Verlag, Berlin, Heidelberg.

Schulz-Mirbach T, Metscher B, Ladich F (2012) Relationship between Swim bladder morphology and hearing abilities-A case study on Asian and African Cichlids. PLoS One 7:e42292. doi: 10.1371/journal.pone.0042292.

Stout JD (1956) Reaction of Ciliates to Environmental Factors. Ecology 37:178–191.

Womack MC, Christensen-Dalsgaard J, Coloma LA, Hoke KL (2018) Sensitive high-frequency hearing in earless and partially eared harlequin frogs (Atelopus). J Exp Biol 221:jeb169664. doi: 10.1242/jeb.169664.

Wörner FG (2015) Schleiereule und Waldkauz Zwei Bewohner der 'Eulenscheune' im Tierpark Niederfischbach. fwö 06:1–28.

Young BA (2003) Snake bioacoustics: toward a richer understanding of the behavioural ecology of snakes. Q Rev Biol 78:303–325.

Chapter 3. Unicellular organisms — communication in the smallest spaces

Bubendorfer S (2013) Flagellen-vermittelte Motilität in Shewanella: Mechanismen zur effektiven Fortbewegung in *S. putrefaciens* CN-32 und *S. oneidensis* MR-1. Doktorarbeit. Philipps-Universität Marburg.

Buonanno F, Harumoto T, Ortenzi C (2013) The Defensive Function of Trichocysts in Paramecium tetraurelia Against Metazoan Predators Compared with the Chemical Defense of Two Species of Toxin-containing Ciliates. Zoolog Sci 30:255–261.

Fritsche O (2016) Mikrobiologie. Springer-Verlag, Berlin, Heidelberg.

Harumoto T, Miyake A (1991) Defensive function of trichocysts in Paramecium. J Exp Zool 260:84–92.

Heitman J (2015) Evolution of sexual reproduction: A view from the fungal kingdom supports an evolutionary epoch with sex before sexes. Fungal Biol Rev 29:108–117.

Horiuchi J, Prithiviraj B, Bais HP, et al (2005) Soil nematodes mediate positive interactions between legume plants and rhizobium bacteria. Planta 222:848–857.

Jarrell KF, McBride MJ (2008) The surprisingly diverse ways that prokaryotes move. Nat Rev Microbiol 6:466–476.

Kirk DL (2004) Volvox. Curr Biol 14:R599-R600.

Lenz P, Hartline DK, Purcell J, Macmillian D (1995) Zooplankton: Sensory Ecology and Physiology. CRC Press.

Liu DWC, Thomas JH (1994) Regulation of a periodic motor program in C. *elegans*. J Neurosci 14:1953–1962.

Magariyama Y, Sugiyama S, Muramoto Y, et al (1994) Very fast flagellar rotation. Nature 371:752.

Matsuhashi M, Pankrushina AN, Takeuchi S, et al (1998) Production of sound waves by bacterial cells and the response of bacterial cells to sound. J Gen Appl Microbiol 44:49–55.

Maynard Smith J (1971) What use is sex? J Theor Biol 30:319–335.

Munk K, Requena N, Fischer R (2008) Taschenlehrbuch Biologie: Mikrobiologie, 2. Auflage. Georg Thieme Verlag, Stuttgart.

Narra HP, Ochman H (2006) Of What Use Is Sex to Bacteria? Curr Biol 16:705–710.

Pandya S, Iyer P, Gaitonde V, et al (1999) Chemotaxis of rhizobium SP.S2 towards *Cajanus cajan* root exudate and its major components. Curr Microbiol 38:205–209.

Sapper N, Widhalm H (2001) Einfache biologische Experimente. Ein Handbuch – nicht nur für Biologen. öbv & hpt, Stuttgart.

Schopf JW, Kitajima K, Spicuzza MJ, et al (2018) SIMS analyses of the oldest known assemblage of microfossils document their taxon-correlated carbon istotope compositions. Proc Natt Acad Sci 115:53 LP–58. doi: 10.1073/ pnas.1718063115.

Silva-Junior EA, Ruzzini AC, Paludo CR, et al (2018) Pyrazines from bacteria and ants: Convergent chemistry within an ecological niche. Sci Rep 8:2595. doi: 10.1038/s41598-018-20953-6.

Troemel ER, Kimmel BE, Bargmann CL (1997) Reprogramming chemotaxis responses: Sensory neurons define olfactory preferences in *C. elegans.* Cell 91:161–169.

Wendel C (2001) Biologische Grundversuche, S I. Bd. 1. Köln.

Werner D (1992) Physiology of nitrogen-fixing legume nodules: compartments and functions. In: Stacy G, Evans HJ, Burris RH (eds) Biological nitrogen fixation. Verlag Chapman and Hall, London, pp. 399–431.

Wheeler JW, Blum MS (1973) Alkylpyrazine alarm pheromones in ponerine ants. Science 182:501–503.

Wicklow BJ (1997) Signal-induced Defensive Phenotypic Changes in Ciliated Protists Morphological and Ecological Implications for Predator and Prey. J Eukaryot Microbiol 44:176–188.

Witzany G (2011) Biocommunication in Soil Microorganisms. Springer Science & Business Media. Heidelberg, London, New York.

Chapter 4. Multicellular organisms — the language of fungi and plants

Adamo SA (1998) Feeding suppression in the tobacco hornworm, Manduca sexta: costs and benefits to the parasitic wasp *Cotesia congregata*. Can J Zool 76:1634–1640.

Allmann S, Baldwin IT (2010) Insects betray themselves in nature to predators by rapid isomerization of green leaf volatiles. Science 329:1075–1078.

Appel HM, Cocroft RB (2014) Plants respond to leaf vibrations caused by insect herbivore chewing. Oecologia 175:1257–1266.

Balan J, Lechevalier HA (1972) The Predaceous Fungus *Arthrobotrys dactyloides*: Induction of Trap Formation. Mycologia 64:919–922.

Baldwin IT, Schultz JC (1983) Rapid changes in tree leaf chemistry induced by damage: evidence for communication between plants. Science 221:277–279

Baluška F, Mancuso S (2018) Plant Cognition and Behavior: From Environmental Awareness to Synaptic Circuits Navigating Root Apices. In: Baluška F, Gagliano M, Witzany G (eds) Memory and Learning in Plants. Springer International Publishing, Cham, pp. 51–77.

Bauer U, Bohn HF, Federle W (2008) Harmless nectar source or deadly trap: Nepenthes pitchers are activated by rain, condensation and nectar. Proc R Soc B Biol Sci 275:259–265.

Billiard S, López-Villavicencio M, Devier B, et al (2011) Having sex, yes, but with whom? Inferences from fungi on the evolution of anisogamy and mating types. Biol Rev 86:421–442.

Bohn HF, Federle W (2004) Insect aquaplaning: Nepenthes pitcher plants capture prey with the peristome, a fully wettable water-lubricated anisotropic surface. Proc Natl Acad Sci 101:14138–14143.

Bouwmeester HJ, Verstappen FWA, Posthumus MA, Dicke M (1999) Spider Mite-Induced (3 S)-(E)-Nerolidol Synthase Activity in Cucumber and Lima Bean. The First Dedicated Step in Acyclic C11-Homoterpene Biosynthesis. Plant Physiol 121:173–180.

Calvo P (2016) The philosophy of plant neurobiology: a manifesto. Synthese 193:1323–1343.

Clarke CM, Kitchings RL (1995) Swimming ants and pitcher plants: A unique ant-plant interaction from Borneo. J Trop Ecol 11:589–602.

de Jager ML, Willis-Jones E, Critchley S, Glover BJ (2017) The impact of floral spot and ring markings on pollinator foraging dynamics. Evol Ecol 31:193–204.

de la Pena C, Badri CD V, Loyola-Vargas V (2012) Plant root secretions and their interactions with neighbors. In: Vivanco J M, Baluška F (eds) Secretions and Exudates in Biological Systems. Springer-Verlag, Berlin, Heidelberg, pp. 1–26.

Elhakeem A, Markovic D, Broberg A, et al (2018) Aboveground mechanical stimuli affect belowground plant-plant communication. PLoS One 13:1–15.

Evans HC, Elliot SL, Hughes DP (2011) *Ophiocordyceps unilateralis*: A keystone species for unraveling ecosystem functioning and biodiversity of fungi in tropical forests? Commun Integr Biol 4:598–602.

Gagliano M (2012) Green symphonies: A call for studies on acoustic communication in plants. Behav Ecol 24:789–796.

Gagliano M, Grimonprez M (2015) Breaking the Silence – Language and the Making of Meaning in Plants. Ecopsychology 7:145–152.

Gagliano M, Grimonprez M, Depczynski M, Renton M (2017) Tuned in: plant roots use sound to locate water. Oecologia 184:151–160.

Gagliano M, Mancuso S, Robert D (2012) Towards understanding plant bioacoustics. Trends Plant Sci 17:323–325.

Gagliano M, Renton M (2013) Love thy neighbour: Facilitation through an alternative signalling modality in plants. BMC Ecol 13:1–6.

Geng S, De Hoff P, Umen JG (2014) Evolution of Sexes from an Ancestral Mating-Type Specification Pathway. PLoS Biol 12:e1001904. doi: 10.1371/journal.pbio.1001904.

Ghergel F, Krause K (2012) Role of Mycorrhiza in Re-forestation at Heavy Metal-Contaminated Sites. In: Bio-geo Interactions in Metal-Contaminated Soils. Springer-Verlag, Berlin, Heidelberg, pp. 183–199.

Heil M, Karban R (2010) Explaining evolution of plant communication by airborne signals. Trends Ecol Evol 25:137–144.

Hughes DP, Wappler T, Labandeira CC (2010) Ancient death-grip leaf scars reveal ant-fungal parasitism. Biol Lett 7:67–70.

Jansson H-B, Nordbring-Hertz B (1979) Attraction of Nematodes to Living Mycelium of Nematophagous Fungi. J Gen Microbiol 112:89–93.

Karban R, Baldwin IT (1997) Induced Responses to Herbivory. University of Chicago Press.

Karban R, Shiojiri K, Ishizaki S, et al (2013) Kin recognition affects plant communication and defence. Proc R Soc B Biol Sci 280:20123062.

Karban R, Yang LH, Edwards KF (2014) Volatile communication between plants that affects herbivory: A meta-analysis. Ecol Lett 17:44–52.

Kessler A, Baldwin IT (2001) Defensive function of herbivore-induced plant volatile emissions in nature. Science 291:2141–2144.

Kothe E (2016) Signalmoleküle in der Mykorrhizasymbiose. In: Die Sprache der Moleküle – Chemische Kommunikation in der Natur. Dr. Friedrich Pfeil, München, pp. 93–103.

Kück U, Wolff G (2014) Botanisches Grundpraktikum. 3. Auflage. SpringerVerlag, Berlin, Heidelberg.

Kullenberg B (1961) Studies in *Ophrys* pollination. Zool Bidr från Uppsala 34:1–340.

Mattiacci L, Dicke M, Posthumus MA (2006) beta-Glucosidase: an elicitor of herbivore-induced plant odor that attracts host-searching parasitic wasps. Proc Natl Acad Sci 92:2036–2040.

Moran JA, Webber EB, Joseph KC (1999) Aspects of Pitcher Morphology and Spectral Characteristics of Six Bornean Nepenthes Pitcher Plant Species: Implications for Prey Capture. Ann Bot 83:521–528.

Nilsson LA (1983) Mimesis of bellflower (Campanula) by the red helleborine orchid *Cephalanthera rubra*. Nature 305:799–800.

Paulus HF (2018) Pollinators as isolation mechanisms: field observations and field experiments regarding specificity of pollinator attraction in the genus *Ophrys* (Orchidaceae und Insecta, Hymenoptera, Apoidea). Entomol Gen 37:261–316.

Qadri AN (1989) Fungi associated with sugarbeet cyst nematode in Jerash, Jordan.

Rhoades DF (1983) Responses of alder and willow to attack by tent caterpillars and webworms: evidence for pheromonal sensitivity of willows. Plant Resist. to insects 208:4–55.

Schaefer HM, Schaefer V, Levey DJ (2004) How plant-animal interactions signal new insights in communication. Trends Ecol Evol 19:577–584.

Schaefer M, Ruxton GD (2004) Communication and the evolution of plant–animal interactions. In: Schaefer HM, Ruxton GD (eds) Plant-Animal Communication. Oxford Scholarship Online, pp. 1–20.

Schenk HJ, Callaway RM, Mahall BE (1999) Spatial Root Segregation: Are Plants Territorial? Adv Ecol Res 28:145–180.

Siddiqui ZA, Mahmood I (1996) Biological control of plant parasitic nematodes by fungi: A review. Bioresour Technol 58:229–239.

Stanley DA, Otieno M, Steijven K, et al (2016) Pollination ecology of *Desmodium setigerum* (Fabaceae) in Uganda; do big bees do it better? J Pollinat Ecol 19:43–49.

Takabayashi J, Sabelis MW, Janssen A, et al (2006) Can plants betray the presence of multiple herbivore species to predators and parasitoids? The role of learning in phytochemical information networks. Ecol Res 21:3–8.

Thornham DG, Smith JM, Ulmar Grafe T, Federle W (2012) Setting the trap: Cleaning behaviour of *Camponotus schmitzi* ants increases long-term capture efficiency of their pitcher plant host, *Nepenthes bicalcarata*. Funct Ecol 26:11–19.

van Dam NM, Bouwmeester HJ (2016) Metabolomics in the Rhizosphere: Tapping into Belowground Chemical Communication. Trends Plant Sci 21:256–265.

Wagner K, Linde J, Krause K, et al (2015) *Tricholoma vaccinum* host communication during ectomycorrhiza formation. FEMS Microbiol Ecol 91:fiv120.

Wells K, Lakim MB, Schulz S, Ayasse M (2011) Pitchers of Nepenthes rajah collect faecal droppings from both diurnal and nocturnal small mammals and emit fruity odour. J Trop Ecol 27:347–353.

Westerkamp C (1997) Keel blossoms: Bee flowers with adaptations against bees. Flora 192:125–132.

Willmer P, Stanley DA, Steijven K, et al (2009) Bidirectional Flower Color and Shape Changes Allow a Second Opportunity for Pollination. Curr Biol 19:919–923.

Wu J, Hettenhausen C, Schuman MC, Baldwin IT (2008) A Comparison of Two *Nicotiana attenuata* Accessions Reveals Large Differences in Signaling Induced by Oral Secretions of the Specialist Herbivore *Manduca sexta*. Plant Physiol 146:927–939.

Chapter 5. Multicellular organisms — communication is animal magic

Alerstam T (1987) Radar observations of the stoop of the Peregrine Falcon *Falco peregrinus* and the Goshawk *Accipiter gentilis*. 129:267–273.

Aquiloni L, Gherardi F (2010) Crayfish females eavesdrop on fighting males and use smell and sight to recognize the identity of the winner. Anim Behav 79:265–269.

Barrett-Lennard LG, Ford JKB, Heise KA (1996) The mixed blessing of echolocation: differences in sonar use by fish-eating and mammal-eating killer whales. Anim Behav 51:553–565.

Bergbauer M (2018) Was lebt in tropischen Meeren? Franckh-Kosmos VerlagsGmbH & Company KG.

Beyer M, Czaczkes TJ, Tuni C (2018) Does silk mediate chemical communication between the sexes in a nuptial feeding spider? Behav Ecol Sociobiol 72:1–9.

Breithaupt T, Eger p. (2002) Urine makes the difference: Chemical communication in fighting crayfish made visible. J Exp Biol 205:1221–1231.

Buchanan KL, Catchpole CK (1997) Female choice in the sedge warbler, *Acrocephalus schoenobaenus*: Multiple cues from song and territory quality. Proc R Soc B Biol Sci 264:521–526.

Burns E, Ilan M (2003) Comparison of anti-predatory defenses of Red Sea and Caribbean sponges. II. Physical defense. Mar Ecol Prog Ser 252:115–123.

Catchpole CK (1980) Sexual selection and the evolution of complex songs among European warblers of the genus Acrocephalus. Behaviour 74:149–165.

Charlton BD, Ellis WAH, Brumm J, et al (2012) Female koalas prefer bellows in which lower formants indicate larger males. Anim Behav 84:1565–1571.

Charlton BD, Frey R, McKinnon AJ, et al (2013) Koalas use a novel vocal organ to produce unusually low-pitched mating calls. Curr Biol 23:1035–1036.

Daura-Jorge FG, Cantor M, Ingram SN, et al (2012) The structure of a bottlenose dolphin society is coupled to a unique foraging cooperation with artisanal fishermen. Biol Lett 8:702–705.

Dean J, Aneshansley DJ, Edgerton HE, Eisner T (1990) Defensive spray of the bombardier beetle: A biological pulse jet. Science 248:1219–1221.

Deecke VB, Slater PJB, Ford JKB (2002) Selective habituation shapes acoustic predator recognition in harbour seals. Nature 420:171.

Deecke VB, Ford JKB, Slater PJB (2005) The vocal behaviour of mammal-eating killer whales: communicating with costly calls. Anim Behav 69:395–405.

Donaghey R (1981) Parental strategies in the green catbird (*Ailuroedus crassirostris*) and the satin bower-bird (*Ptilonorhynchus violaceus*). Monash University, Melbourne, Victoria.

Earley RL, Dugatkin LA (2002) Eavesdropping on visual cues in green swordtail (*Xiphophorus helleri*) fights: A case for networking. Proc R Soc B Biol Sci 269:943–952.

Eckert J (2008) Lehrbuch der Parasitologie für die Tiermedizin. 3. Auflage. Georg Thieme Verlag, Stuttgart.

Ford JKB, Ellis GM, Barrett-Lennard LG, et al (1998) Dietary specialization in two sympatric populations of killer whales (*Orcinus orca*) in coastal British Columbia and adjacent waters. Can J Zool 76:1456–1471.

Francq EN (1969) Behavioral Aspects of Feigned Death in the Opossum *Didelphis marsupialis*. Am Midl Nat 81:556–568.

Freeman AS (2007) Specificity of induced defenses in *Mytilus edulis* and asymmetrical predator deterrence. Mar Ecol Prog Ser 334:145–153.

Frisch K, Chadwick LE (1967) The Dance Language and Orientation of Bees. Harvard Univ. Press, Cambridge, MA.

Gäde G, Weeda E, Gabbott PA (1978) Changes in the Level of Octopine during the Escape Responses of the Scallop, *Pecten maximus* (L.). J Comp Physiol B 124:121–127.

Gewalt W (1965) Formverändernde Strukturen am Halse der männlichen Großtrappe (*Otis tarda* L.). Bonner zool. Beiträge 16:288–300.

Gey MH (2017) Instrumentelles und Bioanalytisches Praktikum. SpringerVerlag, Berlin Heidelberg.

Gorman ML, Mills MGL (1984) Scent marking strategies in hyaenas (Mammalia). J Zool 202:535–547.

Gorman ML (1990) Scent-marking strategies in mammals. Rev Suisse Zool 97:3–29.

Gosling LM, Roberts SC (2001) Testing ideas about the function of scent marks in territories from spatial patterns. Anim Behav 62:F7–F10.

Griffin AS (2004) Social learning about predators: a review and prospectus. Learn Behav 32:131–40.

Hansen LS, Gonzales SF, Toft S, Bilde T (2008) Thanatosis as an adaptive male mating strategy in the nuptial gift-giving spider *Pisaura mirabilis*. Behav Ecol 19:546–551.

Hawkins AD, Johnstone ADF (1978) The hearing of the Atlantic salmon, *Salmo salar*. J Fish Biol 13:655–673.

Haydak MH (1945) The language of the honeybees. Am Bee J 85:316–317.

Herberholz J, Schmitz B (1998) Role of mechanosensory stimuli in intraspecific agonistic encounters of the snapping shrimp (*Alpheus heterochaelis*). Biol Bull 195:156–167.

Hesterman ER, Mykytowycz R (1968) Some observations on the odours of anal gland secretions from the rabbit, *Oryctolagus cuniculus* (L.). CSIRO Wildl Res 13:71–81.

Hidalgo De Trucios SJ, Carranza J (1991) Timing, structure and functions of the courtship display in male great bustard. Ornis Scand 22:360–366.

Hoving HJT, Bush SL, Robison BH (2012) A shot in the dark: Same-sex sexual behaviour in a deep-sea squid. Biol Lett 8:287–290.

Hutchings MR, Service KM, Harris SE (2002) Is population density correlated with faecal and urine scent marking in European badgers (*Meles meles*) in the UK? Mamm Biol 67:286–293.

Irwin MT, Samonds KE, Raharison J, Wright PC (2004) Lemur Latrines: Observations of Latrine Behavior in Wild Primates and Possible Ecological Significance. J Mammal 85:420–427.

Janik VM, Sayigh LS, Wells RS (2006) Signature whistle shape conveys identity information to bottlenose dolphins. Proc Natl Acad Sci 103:8293–8297.

Johansson BG, Jones TM (2007) The role of chemical communication in mate choice. Biol Rev 82:265–289.

Jordan NR, Cherry MI, Manser MB (2007) Latrine distribution and patterns of use by wild meerkats: implications for territory and mate defence. Anim Behav 73:613–622.

Kamio M, Nguyen L, Yaldiz S, Derby CD (2010) How to produce a chemical defense: Structural elucidation and anatomical distribution of aplysioviolin and phycoerythrobilin in the sea hare *Aplysia californica*. Chem Biodivers 7:1183–1197.

Kelley LA, Endler JA (2017) How do great bowerbirds construct perspective illusions? R Soc Open Sci 4:160661. doi: 10.1098/rsos.160661.

Kruuk H (1978) Spatial organization and territorial behaviour of the European badger *Meles meles*. J Zool 184:1–19.

Land BB, Seeley TD (2004) The grooming invitation dance of the honey bee. Ethology 110:1–10.

Lewanzik D, Goerlitz HR (2018) Continued source level reduction during attack in the low-amplitude bat *Barbastella barbastellus* prevents moth evasive flight. Funct Ecol 32:1251–1261.

Lloyd JE (1975) Aggressive Mimicry in Photuris Fireflies: Signal Repertoires by Femmes Fatales. Science 187:452–453.

Lück E, Jager M (2013) Chemische Lebensmittelkonservierung: Stoffe–Wirkungen–Methoden. Springer-Verlag, Heidelberg, Berlin.

MacColl R, Galivan J, Berns DS, et al (1990) The chromophore and polypeptide composition of Aplysia ink. Biol Bull 179:326–331.

MacDonald DW (1980) Patterns of scent marking with urine and faeces amongst carnivore communities. In: Symposia of the Zoological Society of London. pp. 107–139.

Manser MB (2001) The acoustic structure of suricates' alarm calls varies with predator type and the level of response urgency. Proc R Soc B Biol Sci 268:2315–2324.

Marzo V Di, Marin A, Vardaro R, et al (1993) Histological and biochemical bases of defense mechanisms in four species of *Polybranchioidea ascoglossan* molluscs. Mar Biol 117:367–380.

Miller PJO, Shapiro AD, Tyack PL, Solow AR (2004) Call-type matching in vocal exchanges of free-ranging resident killer whales, *Orcinus orca*. Anim Behav 67:1099–1107.

Mills MGL, Gorman ML, Mills MEJ (1980) The scent marking behaviour of the brown hyaena *Hyaena brunnea*. S Afr J Zool 15:240–248.

Milum VG (1955) Honey bee communication. Am Bee J 95:97–104.

Monclús R, de Miguel FJ (2003) Distribución espacial de las letrinas de conejo (*Oryctolagus cuniculus*) en el Monte de Valdelatas (Madrid). Galemys 15:157–165.

Müller WA, Frings S, Möhrlen F (2019) Tier- und Humanphysiologie: Eine Einführung. Springer-Verlag, Berlin, Heidelberg.

Mykytowycz R (1974) Odor in the spacing behaviour of mammals. In: Birch MC (ed) Pheromones. Amsterdam: North-Holland, pp. 327–343.

Mykytowycz R (1962) Territorial Function of Chin Gland Secretion in the Rabbit, *Oryctolagus cuniculus* (L.). Nature 193:799. doi: 10.1038/193799a0.

Mykytowycz R (1968) Territorial marking by rabbits. Sci Am 218:116–126.

Mykytowycz R (1964) Territoriality in rabbit populations. Aust Nat Hist 14:326–329.

Mykytowycz R, Gambale S (1969) The Distribution of Dung-Hills and the Behaviour of free living Wild Rabbits, *Oryctolagus cuniculus* (L.), on them. Forma Funct 1:333–349.

Mykytowycz R, Hestermann ER (1975) An Experimental Study of Aggression in Captive European Rabbits, *Oryctolagus cuniculus*. Behaviour 52:104–123.

Mykytowycz R, Hesterman ER (1970) The behaviour of captive wild rabbits, *Oryctolagus cuniculus* (L.) in response to strange dung-hills. Forma Funct 2:1–12.

Nachtigall W (2013) Biomechanik: Grundlagen, Beispiele, Übungen. Vieweg & Sohn Verlagsgesellschaft mbH, Braunschweig, Wiesbaden.

Nolen TG, Johnson PM, Kicklighter CE, Capo T (1995) Ink secretion by the marine snail *Aplysia californica* enhances its ability to escape from a natural predator. J Comp Physiol A 176:239–254.

Pawlik JR, Chanas B, Toonen RJ, Fenical W (1995) Defenses of Caribbean sponges against predatory reef fish. I. Chemical deterrency. Mar Ecol Prog Ser 127:183–194.

Penney HD, Hassall C, Skevington JH, et al (2012) A comparative analysis of the evolution of imperfect mimicry. Nature 483:461–464.

Pietsch TW, Balushkin A V., Fedorov V V. (2006) New records of the rare deep-sea anglerfish *Diceratias trilobus* Balushkin and Fedorov (Lophiiformes: Ceratioidei: Diceratiidae) from the Western Pacific and Eastern Indian Oceans. J Ichthyol 46:S97–S100.

Prange S, Gehrt SD, Wiggers EP (2003) Demographic Factors Contributing to High Raccoon Densities in Urban Landscapes. J Wildl Manage 67:324–333.

Quaisser C (1996) Der Einfluß von Reizen auf die Herzschlagrate brütender Großtrappen (*Otis t. tarda* L., 1758). Naturschutz und Landschaftspfl Brand 5:103–121.

Quaisser C, Hüppop O (1995) Was stört den Kulturfolger Großtrappe *Otis tarda* in der Kulturlandschaft? Der Ornithol Beobachter 92:269–274.

Reber SA, Townsend SW, Manser MB (2013) Social monitoring via close calls in meerkats. Proc R Soc B Biol Sci 280:20131013. doi: 10.1098/rspb.2013.1013

Ritzmann RE (1974) Mechanisms for the snapping behavior of two alpheid shrimp; *Alpheus californiensis* and *Alpheus heterochelis*. J Comp Physiol 95:217–236.

Roper TJ, Conradt L, Butler J, et al (1993) Territorial marking with faeces in badgers (*Meles meles*): a comparison of boundary and hinterland latrine use. Behaviour 127:289–307.

Roper TJ, Shepherdson DJ, Davies JM (1986) Scent marking with faeces and anal secretion in the European badger (*Meles meles*): seasonal and spatial characteristics of latrine use in relation to territoriality. Behaviour 97:94–117.

Ryne C (2009) Homosexual interactions in bed bugs: alarm pheromones as male recognition signals. Anim Behav 78:1471–1475.

Seyfarth RM, Cheney DL, Marler p. (1980) Monkey responses to three different alarm calls: evidence of predator classification and semantic communication. Science 210:801 LP–803.

Simões-Lopes PC, Fabián ME, Menegheti JO (1998) Dolphin interactions with the mullet artisanal fishing on Southern Brazil: a qualitative and quantitative approach. Rev Bras Zool 15:709–726.

Sneddon IA (1991) Latrine Use by the European Rabbit (*Oryctolagus cuniculus*). J Mammal 72:769–775.

Thomas GE, Gruffydd LD (1971) The types of escape reactions elicited in the scallop *Pecten maximus* by selected sea-star species. Mar Biol 10:87–93.

Thomsen F, Franck D, Ford JKB (2002) On the communicative significance of whistles in wild killer whales (*Orcinus orca*). Naturwissenschaften 89:404–407.

Toledo LF, Sazima I, Haddad CFB (2011) Behavioural defences of anurans: an overview. Ethol Ecol Evol 23:1–25.

Townsend SW, Manser MB (2012) Functionally referential communication in mammals: The past, present and the future. Ethology 118:1–11.

Trussell GC (1996) Phenotypic Plasticity in an Intertidal Snail: The Role of a Common Crab Predator. Evolution (NY) 50:448–454.

Vellenga RETA (1970) Behavior of the male satin bower-bird at the bower. Austral Bird Bander 1:3–8.

Vellenga R (1980) Distribution of bowers of the satin bowerbird *Ptilonorhynchus violaceus*. Emu 81:27–33.

von Byern J, Dorrer V, Merritt DJ, et al (2016) Characterization of the fishing lines in titiwai (=*Arachnocampa luminosa* Skuse, 1890) from New Zealand and Australia. PLoS One 11:e0162687. doi: 10.1371/journal.pone.0162687.

von Holst D, Hutzelmeyer H, Kaetzke P, et al (1999) Social Rank, Stress, Fitness, and Life Expectancy in Wild Rabbits. Naturwissenschaften 86:388–393.

Wickler W (1963) Zum Problem der Signalbildung, am Beispiel der Verhaltens-Mimikry zwischen Aspidontus und Labroides (Pisces, Acanthopterygii). Z Tierpsychol 20:43–48.

Wilson B, Batty RS, Dill LM (2004) Pacific and Atlantic herring produce burst pulse sounds. Proc R Soc B Biol Sci 271:95–97.

Witzany G (2013) Biocommunication of animals. In: Biocommunication of Animals. pp. 1–420.

Wronski T, Plath M (2010) Characterization of the spatial distribution of latrines in reintroduced mountain gazelles (*Gazella gazella*): do latrines demarcate female group home ranges?

Wronski T, Apio A, Plath M, Ziege M (2013) Sex difference in the communicatory significance of localized defecation sites in Arabian gazelles (*Gazella arabica*). J Ethol 31:129–140.

Yeargan KV, Quate LW (1996) Juvenile Bolas Spiders Attract Psychodid Flies. Oecologia 106:266–271.

Yeargan KV (1988) Ecology of a bolas spider, *Mastophora hutchinsoni*: phenology, hunting tactics, and evidence for aggressive chemical mimicry. Oecologia 74:524–530.

Yeargan KV (1994) Biology of Bolas Spiders. Annu Rev Entomol 39:81–99.

Zollner PA, Smith WP, Brennan LA (1996) Characteristics and adaptive significance of latrines of swamp rabbits (*Sylvilagus aquaticus*). J Mammal 77:1049–1058.

Chapter 6. When animals leave the forest

Barrett CG (1901) *B. betularia*. Br Lepid 7:127–134.

Bishop JA (1972) An Experimental Study of the Cline of Industrial Melanism in *Biston betularia* (L.) (Lepidoptera) between urban Liverpool and rural North Wales. J Anim Ecol 41:209–243.

Davison J, Huck M, Delahay RJ, Roper TJ (2009) Restricted ranging behaviour in a high-density population of urban badgers. J Zool 277:45–53.

Defries RS, Foley JA, Asner GP (2004) Land-use choices: balancing human needs and ecosystem function. Front Ecol Environ 2:249–257.

Edleston RS (1864) First carbonaria melanic of moth *Biston betularia*. Entomologist 2:150.

Evans KL, Newton J, Gaston KJ, et al (2012) Colonisation of urban environments is associated with reduced migratory behaviour, facilitating divergence from ancestral populations. Oikos 121:634–640.

Francis RA, Chadwick MA (2012) What makes a species synurbic? Appl Geogr 32:514–521.

Harris S (1982) Activity patterns and habitat utilization of badgers (*Meles meles*) in suburban Bristol: a radio tracking study. In: Symposia of the Zoological Society of London. Published for the Zoological Society by Academic Press, pp. 301–323.

Hof AEV, Campagne P, Rigden DJ, et al (2016) The industrial melanism mutation in British peppered moths is a transposable element. Nature 534:102–105.

Hu Y, Cardoso GC (2009) Are bird species that vocalize at higher frequencies preadapted to inhabit noisy urban areas? Behav Ecol 20:1268–1273.

Johnson MTJ, Munshi-South J (2017) Evolution of life in urban environments. Science 358:eaam8327. doi: 10.1126/science.

Kettlewell HBD (1955) Selection experiments on industrial melanism in the Lepidoptera. Heredity 10:323.

Kettlewell HBD (1958) A survey of the frequencies of *Biston betularia* (L.) (LEP.) and its melanic forms in Great Britain. Heredity 12:51.

LaPoint S, Balkenhol N, Hale J, et al (2015) Ecological connectivity research in urban areas. Funct Ecol 29:868–878.

Luniak M (2004) Synurbanization – adaptation of animal wildlife to urban development. In: Shaw WW, Harris LK, Vandruff L (eds) Proceedings of the 4th International Urban Wildlife Symposium. University of Arizona, Tucson, Arizona, USA, pp. 50–55.

Majerus MEN (2009) Industrial Melanism in the Peppered Moth, *Biston betularia*: An Excellent Teaching Example of Darwinian Evolution in Action. Evol Educ Outreach 2:63–74.

Nemeth E, Brumm H (2009) Blackbirds sing higher-pitched songs in cities: adaptation to habitat acoustics or side-effect of urbanization? Anim Behav 78:637–641.

Nemeth E, Pieretti N, Zollinger SA, et al (2013) Bird song and anthropogenic noise: vocal constraints may explain why birds sing higher-frequency songs in cities. Proc R Soc B Biol Sci 280:2012–2798.

Nisbet EK, Zelenski JM, Murphy SA (2009) The Nature Relatedness Scale. Linking Individuals' Connection With Nature to Environmental Concern and Behavior. Environ Behav 41:715–740.

Prange S, Gehrt SD, Wiggers EP (2003) Demographic Factors Contributing to High Raccoon Densities in Urban Landscapes. J Wildl Manage 67:324–333.

Rabin LA, McCowan B, Hooper SL, Owings DH (2003) Anthropogenic Noise and its Effect on Animal Communication: An Interface Between Comparative Psychology and Conservation Biology. Int J Comp Psychol ISCP 16:172–192.

Rodewald AD, Gehrt SD (2014) Wildlife Population Dynamics in Urban Landscapes. In: McCleery RA, Moorman CE, Peterson MN (eds) Urban Wildlife Conservation – Theory and Praxis. Springer, New York, pp. 117–147.

Roper TJ, Conradt L, Butler J, et al (1993) Territorial marking with faeces in badgers (*Meles meles*): a comparison of boundary and hinterland latrine use. Behaviour 127:289–307.

Roper TJ, Shepherdson DJ, Davies JM (1986) Scent marking with faeces and anal secretion in the European badger (*Meles meles*): seasonal and spatial characteristics of latrine use in relation to territoriality. Behaviour 97:94–117.

Russell R, Guerry AD, Balvanera P, et al (2013) Humans and Nature: How Knowing and Experiencing Nature Affect Well-Being. Annu Rev Environ Resour 38:473–502.

Ryan AM, Partan SR (2014) Urban Wildlife Behavior. In: Urban Wildlife Conservation – Theory and Praxis. pp. 149–173.

Šálek M, Drahníková L, Tkadlec E (2015) Changes in home range sizes and population densities of carnivore species along the natural to urban habitat gradient. Mamm Rev 45:1–14.

Slabbekoorn H (2013) Songs of the city: noise-dependent spectral plasticity in the acoustic phenotype of urban birds. Anim Behav 85:1089–1099.

Slabbekoorn H, Boer-Visser A den (2006) Cities Change the Songs of Birds. Curr Biol 16:2326–2331.

Slabbekoorn H, Peet M (2003) Birds sing at a higher pitch in urban noise. Nature 424:267.

Tucker MA, Böhnung-Gaese K, Fagan WF, et al (2018) Moving in the Anthropocene: Global reductions in terrestrial mammalian movements. Science 359:466–469.

Tutt JW (1896) British moths. George Routledge, London.

Vining J, Merrick MS, Price EA (2008) The Distinction between Humans and Nature: Human Perceptions of Connectedness to Nature and Elements of the Natural and Unnatural. Hum Ecol Rev 15:1–11.

Wiley RH, Richards DG (1978) Physical constraints on acoustic communication in the atmosphere: implications for the evolution of animal vocalizations. Behav Ecol Sociobiol 3:69–94.